The Intelligent Life

Words, Manners, Judgment, and the Inner Discipline That Decide Your Future

The Intelligent Life

Words, Manners, Judgment, and the Inner Discipline That Decide Your Future

(Illustrated, Refined & Edited)

Carl Sagon

Beyond Books Hub

Beyond Books Hub

Copyright Notice for Readers

Publication Year: **2026** | Publisher: **Beyond Books**
ISBN: **9789377077143**

Edition: **First Edition, 2026** | Contact: ddsalesglobal@gmail.com

Book Name: **The Intelligent Life**
Author: **Carl Sagon**
Price: **Rs 250/-**

Table of Content

Reader's Note

This book does not rush toward you with loud promises or exaggerated claims. Instead, it approaches quietly—much like the forces that truly shape a life. The Intelligent Life is built on a simple but often ignored truth: most destinies are not destroyed by dramatic failures, but by small, repeated lapses in judgment, discipline, speech, and self-awareness. Likewise, most successes are not accidents—they are carefully, patiently constructed from the inside outward.

Across these chapters, the reader is invited into a thoughtful exploration of everyday moments that usually pass unnoticed: how we argue, how we control our impulses, how we speak, read, relax, and interact with others. These moments may seem ordinary, but together they form the hidden structure of character. This book treats them with the seriousness they deserve—without becoming heavy or moralistic.

Each chapter offers more than advice; it offers perspective. The Quiet Power of Purposeful Argument teaches that intelligence is not proven by volume, but by clarity and restraint. Mastering the Inner Fire turns emotional energy from a liability into a source of strength. The Quiet Trap of Being Fooled sharpens awareness in a world full of influence, persuasion, and half-truths. As the chapters unfold, readers begin to recognize patterns in their own thinking and behavior—not through accusation, but through insight.

What makes this book especially engaging is its balance between intellect and accessibility. The themes are deep, yet the language is clear. The ideas are serious, yet the reading experience remains fluid and absorbing. Discussions on manners, conversation, leisure, and reading habits are not presented as outdated rules, but as living skills that determine how one is perceived, trusted, and remembered in society.

There is also a quiet encouragement running through every page. This is not a book that talks down to its reader. It assumes intelligence, curiosity, and the desire to grow. It respects the reader's capacity to reflect and improve. Whether you are a student shaping your future, a professional refining your conduct, or a lifelong learner seeking clarity, this book meets you where you are—and gently moves you forward.

Most importantly, The Intelligent Life connects self-discipline with dignity, manners with self-respect, and reading with influence. It reminds us that intelligence is not only what we know, but how we live. By the time you finish this book, you may not feel dramatically transformed—but you will think more clearly, observe more carefully, and act more deliberately. And that is precisely how real change begins.

Publisher's Note

At a time when speed is celebrated more than substance, and expression often outweighs reflection, The Intelligent Life stands as a thoughtful counterbalance. From the beginning, our goal as publishers was not to modernize the message by altering its meaning, but to present it in a form that today's readers can engage with easily, thoughtfully, and fully.

This book addresses timeless concerns: judgment, discipline, speech, manners, learning, and social conduct. These are not trends bound to a particular era; they are enduring foundations of personal and social success. However, language, reading habits, and attention spans evolve. Our responsibility, therefore, was to ensure that the presentation of these ideas remains clear and inviting—without compromising their depth or seriousness.

In preparing this edition, we focused on clarity, flow, and readability. Complex ideas have been allowed to breathe. Sentences are structured to guide rather than overwhelm. The tone remains respectful and reflective, preserving the original seriousness of the subject while ensuring that modern readers can move through the book with ease and interest.

We believe deeply in books that develop the individual, not merely entertain for the moment. The Intelligent Life belongs to that tradition of works meant to be revisited, reflected upon, and gradually absorbed. It does not demand agreement; it invites consideration. It does not prescribe rigid formulas; it offers principles that readers can apply thoughtfully to their own circumstances.

As publishers, we are committed to producing works that honor the intelligence of the reader. This means resisting simplification that dilutes meaning, while also avoiding unnecessary complexity that creates distance. The present edition reflects that balance. It respects the author's vision and intent while acknowledging the realities of contemporary reading.

We offer this book to readers who value substance over spectacle, growth over shortcuts, and understanding over noise. Our hope is that this edition serves not only as a book to be read, but as a quiet companion in the ongoing process of self-education and personal refinement.

Introduction

A life is rarely shaped by one defining moment. It is shaped instead by hundreds of small decisions—how we respond when challenged, how we speak when emotions rise, how we use our time when no one is watching, and how seriously we take the responsibility of learning. The Intelligent Life is a journey into those subtle but powerful forces that determine who we become.

From the very first chapter, the reader is drawn into an examination of inner strength—the kind that does not announce itself, but steadily influences outcomes. Purposeful argument, emotional control, and intellectual alertness are presented not as abstract ideals, but as practical skills that quietly separate those who drift from those who direct their lives.

As the book progresses, the focus widens. Words are revealed as instruments that shape identity and destiny. Manners are shown not as social decorations, but as indicators of self-respect and awareness. Conversation becomes an art—one that builds connection, trust, and influence when practiced with intelligence and care.

There is also a thoughtful exploration of time—especially the time often wasted or underestimated. Leisure, when used intelligently, becomes restorative rather than destructive. Reading transforms from a passive habit into a deliberate act of selfconstruction. Study is reframed not as pressure, but as discipline that brings freedom, clarity, and confidence.

What gives this book its quiet power is its psychological insight. It recognizes how easily people are misled—by others, by circumstances, and by their own unchecked impulses. Rather than dramatizing this vulnerability, the book treats it as a shared human condition and offers awareness as the remedy.

Throughout the chapters, there is an underlying sense of suspense—not the suspense of external events, but the inner suspense of self-discovery. Readers begin to anticipate moments of recognition, where a sentence reflects a long-felt but unspoken truth. Emotional depth arises not from storytelling, but from realization.

The Intelligent Life does not promise an effortless transformation. It promises something far more reliable: understanding. And from understanding comes direction. This book is for readers who sense that intelligence is more than information, that success is more than achievement, and that a meaningful life is built—quietly, patiently, and deliberately.

As you begin reading, you are not entering a lecture, but a conversation—one that respects your intelligence and challenges you to refine it. The journey ahead is not loud, but it is lasting.

Chapter-1.
The Quiet Power of Purposeful Argument

Not every argument needs to be loud to be powerful. This chapter uncovers how calm, structured reasoning can create clarity instead of conflict. It teaches the value of thoughtful disagreement, helping you transform arguments into opportunities for understanding, influence, and stronger personal conviction rooted in logic rather than emotion.

There is a moment in nearly every life when disagreement stops being noise and begins to matter. It may arrive in a classroom discussion, a family decision, a workplace meeting, or a casual conversation that suddenly turns serious. In that moment, argument reveals its true nature—not as a quarrel or contest of egos, but as a human tool with the power to shape understanding, influence direction, and quietly alter the course of events. To argue effectively is not to win by force of voice or clever insult; it is to participate in a disciplined exchange of thought where reason, listening, and intention work together toward something better than confusion. Many people grow up believing that argument is either pointless or dangerous. They see raised voices, broken friendships, stubborn pride, and assume disagreement itself is the problem. Others swing to the opposite extreme and argue about everything, mistaking contradiction for intelligence and noise for strength. Both views miss the deeper truth. Argument is neither inherently harmful nor automatically useful. Its value depends entirely on how, why, and about what it is used. When argument is guided by purpose, knowledge, and respect, it becomes one of the most powerful instruments of personal growth and collective progress.

Effective argument begins long before words are spoken. It starts with judgment—the ability to recognize which matters deserve argument and which do not. Facts that can be verified with a glance rarely benefit from debate. Matters of personal taste, deeply rooted in emotion or habit, seldom change under pressure. Arguing about such things drains energy without producing insight. Purposeful argument lives in a narrower but richer space: opinions shaped by incomplete information, and proposed actions whose outcomes are still open. In these spaces, argument has room to work. Here, minds can change, decisions can improve, and errors can be corrected before they harden into consequences.

Once a worthy subject is chosen, the character of the argument matters more than the cleverness of the speaker. Knowledge stands first among all requirements. To argue without understanding is not bravery; it is recklessness. Information does not guarantee persuasion, but ignorance guarantees failure. Knowing a subject means more than memorizing points that support one's position. It means understanding the background, the constraints, the risks, and the reasonable objections. A well-informed arguer does not fear questions, because questions reveal where clarity is needed. Informed argument carries a quiet confidence, not because it shouts louder, but because it stands on solid ground.

Yet knowledge alone is insufficient. Many arguments collapse not from lack of facts, but from lack of listening. True listening is an active discipline. It requires setting aside the impulse to interrupt, correct, or rehearse one's reply while the other person speaks. Until an arguer understands what the other side actually believes—not what is assumed, exaggerated, or feared—there is nothing meaningful to respond to. Listening transforms argument from parallel monologues into genuine exchange. It reveals not only disagreement, but also unexpected agreement, shared values, and common goals that can become bridges instead of barriers.

Finding common ground is not a weakness; it is a strategic and human necessity. Two people who share nothing cannot persuade each other. Agreement, even on small points, creates a foundation on which trust and clarity can rest. When common ground is acknowledged, the tone of argument shifts.

It becomes less about opposition and more about problem-solving. From that shared base, differences can be examined without hostility, and disagreement becomes a joint exploration rather than a personal attack.

Clarity emerges when the real issues are identified. Many arguments fail because participants fight shadows instead of substance. They argue around the edges, confuse symptoms with causes, or allow emotions to drag the discussion away from its core. Effective argument requires the discipline to ask: Where, exactly, do we disagree? What assumption differs? What outcome do we fear or hope for? Once the true point of disagreement is named, effort can be focused where it matters. Energy is no longer wasted on side issues or misunderstandings. The argument gains direction, and direction gives it force.

Respect is the quiet structure holding all these elements together. Without respect, knowledge becomes arrogance, listening becomes performance, and common ground becomes manipulation. Respect does not mean agreement, nor does it require softness. It means recognizing the other person as a thinking human being whose views, even when mistaken, arise from reasons that deserve examination. Respect keeps argument from descending into insult and keeps disagreement from becoming personal warfare. It allows firmness without cruelty and conviction without contempt.

Emotion, often blamed for ruining arguments, is not the enemy it is made out to be. Emotion provides motivation, urgency, and human meaning. The problem arises when emotion replaces reasoning instead of supporting it. Anger can signal injustice; enthusiasm can inspire action; concern can sharpen focus. Effective argument acknowledges emotion without surrendering control to it. The skilled arguer recognizes emotional currents—both their own and others'—and steers the discussion without denying its human reality.

Elements of Effective Argument

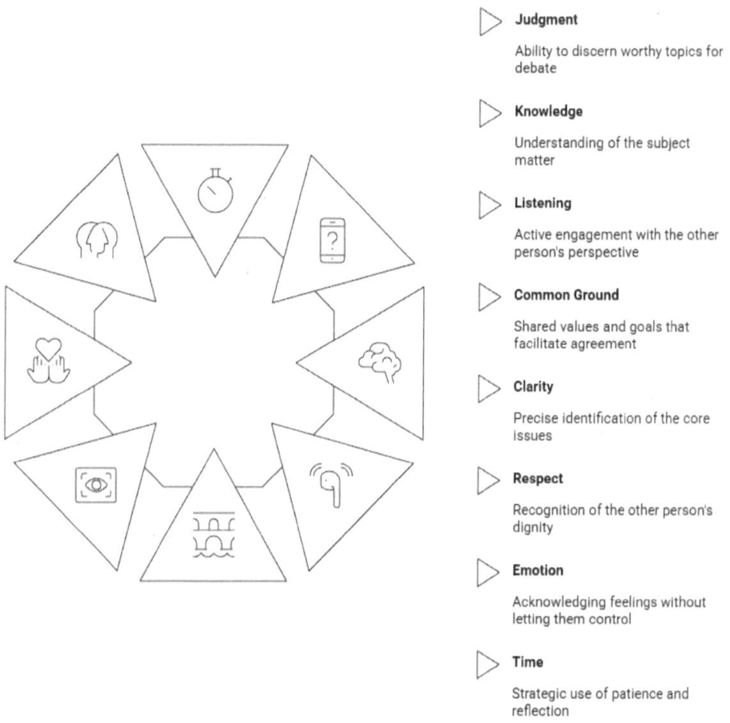

Time also plays a subtle role in argument. Not every disagreement must be resolved immediately. Some arguments require patience, reflection, and the willingness to return later with better information or calmer minds. Rushing an argument often hardens positions instead of softening them. Purposeful argument respects timing, knowing when to press forward and when to pause. This patience is not avoidance; it is strategic restraint.

As one grows in the practice of effective argument, a change occurs that is almost invisible. Arguments become fewer, but more meaningful. The impulse to contradict every questionable statement fades, replaced by a sharper sense of when disagreement truly matters. The arguer begins to conserve energy for issues where influence is possible and consequences are real. This selectivity is not indifference; it is maturity.

Over time, argument reshapes the arguer. Listening improves not only debates, but relationships. Seeking common ground strengthens cooperation. Clarifying issues sharpens thinking in all areas of life.

Even losing an argument becomes valuable, revealing weaknesses in reasoning or gaps in knowledge that can be addressed. Effective argument, practiced honestly, becomes a form of education that never ends.

In communities, workplaces, and societies, the absence of effective argument is far more dangerous than its presence. When people stop arguing thoughtfully, decisions drift toward habit, authority, or impulse. Poor ideas survive unchallenged, and better ones remain unspoken.

Healthy disagreement acts as a corrective force, testing assumptions and refining plans before they harden into action. Argument, when done well, is not a sign of division but of engagement.

The most powerful arguments often appear modest on the surface. They do not rely on dramatic language or theatrical gestures.

They proceed step by step, building understanding, inviting response, and adjusting course when new information appears. Their strength lies not in dominance but in durability. They leave behind not resentment, but clarity.

To learn to argue effectively is, in a quiet way, to learn how to live more deliberately. It teaches patience in a hurried world, listening in a noisy one, and humility in an age that rewards certainty over understanding.

It trains the mind to distinguish between what feels right and what can be shown to be right, between winning a moment and improving a decision.

In the end, effective argument is not about defeating others. It is about engaging reality together—testing ideas, refining plans, and choosing actions with open eyes. Those who master this art do not stop being arguers. They become something better: thoughtful participants in the ongoing conversation that shapes their lives and the lives of those around them.

Questions & Answers

Question 1: How does the judgment to remain silent on certain issues increase an arguer's overall influence?
Answer: Purposeful argument requires the ability to recognize which subjects are worth one's time and energy. Issues based on easily verifiable facts or matters of personal taste rarely benefit from debate; arguing over them only drains energy and credibility. By practicing strategic restraint—choosing carefully when to engage—an arguer speaks less often but with greater impact. Their arguments carry more weight because they are reserved for situations where influence is possible and real consequences are at stake.

Question 2: Why is understanding the constraints and risks of one's own position considered essential for standing on "solid ground"?
Answer: True knowledge goes beyond collecting points that support one's own side. It requires understanding the background, limitations, risks, and reasonable objections related to a position. This broader awareness creates a quiet confidence: the arguer is not caught off guard by difficult questions but welcomes them as opportunities to refine understanding. Such preparedness ensures that arguments are stable, realistic, and credible.

Question 3: In what way does active listening function as a strategic necessity rather than merely a social courtesy?
Answer: Active listening is a disciplined skill that prevents arguments from turning into parallel monologues. Strategically, it is impossible to respond meaningfully without first understanding what the other person truly believes, rather than reacting to assumptions or fears. Listening also reveals areas of

common ground—small agreements that build trust and allow the discussion to shift from hostility toward cooperative problem-solving.

Question 4: How can emotion be integrated into a purposeful argument without replacing reasoning?
Answer: Emotion does not inherently weaken an argument; when guided properly, it adds urgency, motivation, and human significance. Anger can highlight injustice, while concern can sharpen attention and focus. The skilled arguer acknowledges emotional undercurrents and uses them as signals or guides, while ensuring that logic and clarity remain in control of the discussion.

Question 5: Why is the absence of effective argument considered more dangerous to society than its presence?
Answer: When thoughtful disagreement disappears, decisions begin to drift toward habit, authority, or impulse. Without the corrective force of healthy argument, weak ideas go unchallenged and persist, while stronger ideas may never surface. In this sense, argument is not a sign of social division but a vital sign of engagement—a process through which assumptions are tested before they harden into permanent actions.

End of the Chapter

Chapter–2.
Mastering the Inner Fire

Inside you exists a force that can either drive success or lead to chaos. This chapter explores how to control and direct your inner energy—your emotions, ambition, and impulses—so they serve a purpose. It reveals how mastering this fire leads to focus, resilience, and meaningful personal growth.

Before humanity learned how to guide fire, it was a force of terror. It burned forests, destroyed shelters, and threatened survival. Yet the same force, once understood and controlled, became the foundation of progress. Fire warmed homes, cooked food, forged tools, and extended the boundaries of civilization. Human emotion is much the same.

Left uncontrolled, it can harm relationships, distort judgment, and exhaust the body and mind.

Understood and guided, it becomes energy, motivation, warmth, and creative power. Learning to control emotions is not about suppressing feeling or becoming cold and mechanical; it is about mastering a force that already lives within us so that it works for us rather than against us.

Emotions are not accidental flaws in human design. They are deeply woven into our biology and psychology, shaped by millions of years of survival. Fear protects us from danger. Anger mobilizes us against threat or injustice. Love binds us together and gives life meaning. These emotional responses arise quickly, often before conscious thought, because they once needed to. The problem is not that emotions exist, but that modern life constantly stimulates them in ways our nervous systems were never designed to manage without reflection. Traffic delays, social judgments, digital messages, competitive pressures, and unending demands create a steady stream of emotional triggers. Without understanding and control, these triggers accumulate until emotion overwhelms reason.

To control emotions, one must first abandon the illusion that emotions simply "happen" without cause. Every emotional reaction follows a pattern, even when it feels sudden or irrational. Something occurs, the mind interprets it, and the body responds. This is the basic rhythm of emotional life.

A comment, an obstacle, a disappointment, or a perceived threat acts as a stimulus. The mind evaluates it—often unconsciously—based on past experience, beliefs, and expectations. The result is an emotional response that prepares the body for action. This sequence unfolds in fractions of a second, but it is not beyond influence. Control begins with recognizing that emotions are responses, not commands.

At birth, human beings are equipped with a small set of fundamental emotional patterns.

Frustration produces anger when movement or desire is blocked. Sudden danger or instability produces fear. Warmth, care, and closeness produce affection. These primary responses are simple and universal. As life progresses, these basic emotions become associated with countless situations through learning and experience.

A raised voice may evoke fear because it once preceded punishment. A certain tone may evoke anger because it once accompanied disrespect. A familiar gesture may evoke comfort because it once meant safety. Over time, emotional responses become complex, layered, and highly personal.

This learning process is useful, but it also creates vulnerability. When emotional reactions become automatic habits, they may no longer serve present reality. An adult may react with the intensity of childhood fear to situations that are no longer dangerous. A minor inconvenience may trigger rage because it resembles earlier frustrations. Emotional control, therefore, is not about eliminating emotion, but about updating it—bringing response back into alignment with present reality.

The Cycle of Emotional Mastery

A balanced personality does not lack emotion; it integrates it. Emotional balance allows feeling without being ruled by feeling. When emotions are out of control, they narrow attention, distort thinking, and reduce efficiency. Under strong emotional stress, people forget simple facts, make careless mistakes, and act in ways they later regret. The phrase "I couldn't think straight" is not an exaggeration; intense emotion literally interferes with cognitive function. The brain shifts resources from reasoning to reaction, prioritizing survival over judgment. This may be useful in physical danger, but it is destructive in daily life.

Consider how easily small frustrations accumulate. One irritation alone may be manageable, but when several occur in succession, emotional tolerance erodes. A failed attempt, a delay, an unexpected obstacle—each adds pressure. Without awareness, the emotional system reacts not only to the present stimulus but to the entire accumulated load. The result is an overreaction that seems disproportionate, even to the person experiencing it. This is why emotional outbursts often surprise both observers and the individual involved. The final trigger was small; the emotional explosion was fueled by everything that came before.

Another danger of uncontrolled emotion is misdirected blame. When emotional tension builds, the mind searches for an outlet. Objects, circumstances, or people become targets, even when they are not responsible. This process, known informally as blaming the nearest cause, provides temporary relief but solves nothing. It damages relationships and reinforces the habit of emotional irresponsibility. Over time, this pattern becomes automatic: frustration seeks a target, and anger finds one, regardless of accuracy.

Emotions also spread. Human beings are social creatures, highly sensitive to the emotional states of those around them. Anger provokes anger, anxiety provokes anxiety, calm invites calm. This emotional contagion is subtle but powerful. One person's loss of control can ripple outward, altering the mood of a family, a workplace, or a group. This is why emotional discipline is not only a personal matter; it is a social responsibility. Controlling one's emotions protects not only oneself but the emotional well-being of others.

Mixed emotions further complicate matters. Love, fear, and anger can arise together, pulling behavior in conflicting directions. Care for someone may coexist with anger at their actions and fear for their safety. Without control, these mixed emotions can produce chaotic behavior, where actions contradict intentions. Understanding emotional complexity helps prevent this confusion. When people recognize that multiple emotions can coexist, they gain space to choose which one guides action.

Uncontrolled emotional episodes leave lasting traces. Words spoken in anger cannot be unsaid. Actions taken in fear cannot always be undone. Even when apologies are offered, memory remains. Repeated loss of control can gradually shape personality, reinforcing impulsiveness, resentment, or withdrawal. Emotional habits, like physical ones, strengthen through repetition. The more often emotions are allowed to dominate behavior, the harder control becomes.

Fortunately, emotional control is learnable. It is not a fixed trait granted to a lucky few. It develops through understanding, practice, and conscious choice. There are three fundamental ways to influence emotional reactions: altering the situation, altering the meaning of the situation, or altering the response itself.

The first approach is eliminating the stimulus. This means avoiding situations that reliably trigger destructive emotional reactions, especially when no benefit is gained from exposure. If certain environments, habits, or interactions repeatedly provoke loss of control, wise adjustment is not weakness but intelligence. However, life does not always allow avoidance. Many emotional triggers are unavoidable: unexpected failures, other people's behavior, or circumstances beyond control.

When elimination is impossible, modification becomes essential. Modifying the stimulus does not mean denying reality; it means changing perspective. How an event is interpreted determines its emotional impact.

A mistake can be viewed as humiliation or as information. A delay can be seen as an insult or as inconvenience. By consciously reinterpreting situations, emotional intensity can be reduced. This cognitive flexibility is a cornerstone of emotional control. It requires stepping back mentally, observing oneself, and asking whether the emotional meaning assigned is accurate or useful.

The most powerful method, however, is modifying the response. Even when a stimulus remains and interpretation is difficult to change, behavior can still be chosen. Emotion urges action, but it does not dictate it. With practice, people can experience anger without acting aggressively, fear without fleeing, and disappointment without despair. This does not suppress emotion; it channels it. Humor can replace irritation. Patience can replace haste. Curiosity can replace defensiveness. Over time, these alternative responses become habits, just as uncontrolled reactions once were.

Emotional control improves thinking. When emotional intensity is regulated, the mind regains clarity. Problems are approached logically rather than impulsively. Solutions emerge where previously there was only frustration. This clarity reduces mistakes and increases effectiveness. People who manage emotions well often appear calm under pressure, not because they feel less, but because they think through feeling instead of being swept away by it.

Understanding emotions also improves empathy. When individuals recognize their own emotional patterns, they become more sensitive to those of others. They pause before reacting to someone else's anger, recognizing it may be the result of accumulated stress rather than personal hostility. This understanding prevents unnecessary conflict and fosters cooperation. Emotional control, therefore, is not emotional isolation; it is emotional intelligence.

Practice strengthens control. Each time an emotional reaction is guided rather than indulged, the next instance becomes easier. Neural pathways associated with self-regulation grow stronger. What once required effort becomes more natural. This is why emotional maturity often increases with age—experience provides repeated opportunities to practice control. Yet age alone is not enough; awareness is required. Without reflection, years merely reinforce old habits.

Controlling emotions does not mean eliminating passion or enthusiasm. On the contrary, controlled emotion is more effective emotion. Passion directed by purpose achieves far more than passion unleashed without direction. Anger disciplined by reason becomes moral courage. Fear tempered by judgment becomes caution. Love guided by understanding becomes commitment rather than dependence.

A life governed by uncontrolled emotion is exhausting. It drains energy through constant reaction and regret. A life guided by emotional control conserves strength for what truly matters. Decisions improve, relationships deepen, and inner stability grows. Happiness becomes less dependent on external circumstances and more rooted in inner balance.

Ultimately, emotional control is an act of self-respect. It is the decision not to surrender one's behavior to every passing impulse. It is the recognition that while emotions arise automatically, actions need not. By understanding emotional patterns, recognizing triggers, and practicing thoughtful response, individuals reclaim authorship of their lives.

The inner fire never disappears. It should not. But when it is understood and guided, it warms rather than burns. It lights the path forward instead of destroying what lies behind. In mastering emotion, a person does not become less human, but more fully so—capable of feeling deeply, acting wisely, and living with steadiness in an unsteady world.

Questions & Answers

Question 1: Why does a minor inconvenience sometimes trigger a seemingly disproportionate emotional explosion?
Answer: Emotional tolerance often erodes when small frustrations—such as delays, interruptions, or repeated obstacles—occur one after another. The resulting outburst is usually not caused by the final minor trigger alone, but by the **entire accumulated load** of unprocessed irritations. This explains why people are often surprised by the intensity of their own reactions: the last event merely serves as an outlet for a reservoir of pressure that has been building over time.

Question 2: How does the brain's "survival prioritization" sabotage modern decision-making?
Answer: During moments of intense emotion, the brain treats modern stressors—such as digital messages, social pressure, or competition—as if they were physical threats. This triggers a survival response that shifts mental resources away from reasoning and toward rapid reaction. While this mechanism was useful in primitive environments, in everyday life it narrows

attention, distorts judgment, and leads to impulsive actions, careless mistakes, and words that cannot be taken back.

Question 3: In what way is emotional control an act of self-respect?
Answer: Allowing behavior to be ruled by passing impulses is a surrender of personal authorship over one's life. Emotional control represents a deliberate refusal to let every internal feeling or external provocation dictate action. By exercising restraint, an individual reclaims agency and ensures that behavior remains aligned with long-term values and intentions rather than being driven by fleeting, chaotic emotions.

Question 4: Why is modifying the meaning of a situation more sustainable than eliminating the stimulus?
Answer: While avoiding emotional triggers can be intelligent, many triggers—such as failure, uncertainty, or other people's behavior—are unavoidable. A more powerful and sustainable strategy is **cognitive flexibility**: the ability to pause, step back, and reinterpret the situation. By changing the meaning assigned to an event, emotional intensity is reduced at its source, allowing individuals to remain effective and composed even in high-stress environments.

Question 5: What is the relationship between emotional maturity and "updating" one's responses?
Answer: Many emotional reactions in adulthood are automatic habits formed in childhood that no longer fit present reality. Emotional maturity involves recognizing these outdated patterns and consciously adjusting responses to match the current situation. This process of "updating" strengthens neural pathways associated with self-regulation, gradually transforming calm, measured responses into a stable and enduring habit.

End of the Chapter

Chapter-3.
The Quiet Trap of Being Fooled

The most dangerous deceptions are the ones you never notice. This chapter exposes how easily perception can be distorted by assumptions, biases, and misplaced trust. It guides you to sharpen awareness, question appearances, and protect yourself from subtle errors in judgment that can shape your decisions and outcomes.

There is a subtle instant in human life when thought begins to drift from reflection to repetition. It does not announce itself with warning, nor does it arrive as an obvious mistake. It enters quietly, wrapped in familiar language, appealing emotions, and reassuring certainty. A person still feels independent, still believes they are choosing freely, yet something essential has shifted. Ideas are no longer examined; they are accepted because they sound right, because they echo what others say, or because they offer comfort in moments of confusion. This chapter explores that fragile moment when awareness weakens and how, without vigilance, even well-meaning people can be guided into surrendering their freedom while believing they are protecting it.

Freedom is rarely destroyed by force alone. It erodes through habit. It weakens when people stop questioning, when they allow anger to replace understanding and loyalty to replace judgment. Societies do not fall because everyone suddenly agrees to be unfree. They falter because individuals begin to accept small distortions as harmless, small exclusions as necessary, and small lies as practical. Each concession feels temporary and justified. Together, they create a permanent loss that no single person intended.

The danger is not confined to one era, one nation, or one ideology. It is a recurring human vulnerability. Whenever people feel uncertain about their future, anxious about change, or threatened by complexity, the temptation for simple explanations grows strong. Complex systems are difficult to understand. Economic shifts, social change, and cultural diversity require patience and learning.

In moments of stress, many people crave clarity, even if it is false. They want someone to tell them who is responsible and how to fix it quickly. This craving creates opportunity—not for truth, but for manipulation.

Those who exploit this vulnerability rarely present themselves as manipulators. They adopt the language of concern and protection. They speak of tradition, security, identity, and belonging. They frame themselves as defenders of ordinary people against unseen threats. Their message is designed not to inform, but to inflame. They do not ask listeners to think; they ask them to feel. Anger, fear, and resentment become tools. When emotion is activated, reasoning slows, and critical thought recedes.

One of the most effective tools in this process is division. A population that sees itself as a single community is difficult to mislead. Shared experience encourages empathy, cooperation, and resistance to unjust authority.

But when people are separated into categories—when "us" is defined against "them"—unity fractures. Distrust replaces cooperation. Suspicion replaces dialogue. Each group becomes preoccupied with defending itself, leaving little energy to **Question** those who benefit most from the division.

Prejudice thrives in this environment. It offers a ready-made explanation for frustration. It tells people that their struggles are not the result of complex systems, poor leadership, or structural inequality, but of other people who are different from them. This explanation is appealing because it removes responsibility and simplifies reality. It does not require learning, adaptation, or self-reflection. It only requires belief.

Yet prejudice is never accidental. It is cultivated. It is reinforced through repetition, selective storytelling, and emotional language. Over time, it becomes familiar, and familiarity is often mistaken for truth. What begins as suggestion hardens into conviction. People begin to defend these ideas passionately, not because they are accurate, but because they are now tied to identity. Challenging the idea feels like a personal attack.

Once this stage is reached, freedom becomes fragile. Rights begin to feel conditional. People start to accept the idea that some individuals deserve fewer protections, fewer opportunities, or less dignity.

This acceptance is always justified as exceptional, temporary, or necessary. But once the principle is established that rights can be selectively applied, no boundary remains secure. The logic that excludes one group today can be applied to another tomorrow.

History demonstrates this pattern repeatedly, but its lessons are often ignored because each generation believes itself immune. People assume manipulation is obvious, that they would recognize it immediately.

They imagine that only the uneducated or the malicious fall for such tactics. In reality, manipulation works precisely because it appeals to ordinary concerns and emotions. It does not require ignorance; it requires trust without scrutiny.

Silence plays a critical role in this process. When the first group is targeted, many remain quiet, convincing themselves it does not concern them. Silence feels safe. It feels prudent. It feels like survival.

But silence is never neutral. It signals permission. It teaches manipulators how far they can go. Each unchallenged act of exclusion becomes a precedent for the next.

Those who remain silent often believe they are avoiding conflict. In truth, they are postponing it, allowing it to grow stronger and more dangerous. When the circle of exclusion expands and reaches them, they discover that the protections they assumed would remain are gone. By then, the social bonds that might have resisted have already been weakened.

Manipulation also targets institutions that encourage cooperation and shared power.

Any organization that allows people to unite around common interests becomes suspect. Collective action is portrayed as dangerous, disloyal, or corrupt. Individuals are encouraged to distrust one another and rely instead on centralized authority. Isolation increases dependence, and dependence reduces resistance.

Truth becomes inconvenient in such an environment. Independent thinking is framed as arrogance. Expertise is dismissed as elitism. Education is portrayed as indoctrination rather than inquiry. When truth threatens control, it must be discredited or drowned out. The goal is not to prove lies correct, but to make truth irrelevant.

Once this happens, repetition replaces evidence. Messages are repeated until they feel natural. Contradictory information is labeled hostile or fake. People retreat into echo chambers where beliefs are constantly reinforced. Choice becomes an illusion, because alternatives are no longer visible or trusted.

The personal cost of this process is profound. Relationships strain under ideological pressure. Friends become enemies. Families fracture. Fear becomes a constant companion. Even those who believe they benefit from the division live with anxiety, always alert for new threats and new targets. Hatred requires maintenance. It never rests.

The greatest tragedy is that those who are manipulated often believe they are strong. They are told they belong to something greater, something pure and powerful. This sense of belonging feels empowering, but it is hollow. It depends entirely on obedience and conformity. Individual thought becomes dangerous. Questioning becomes betrayal. The promised strength belongs not to the believer, but to those directing belief.

Freedom, by contrast, is demanding. It requires tolerance of uncertainty, respect for difference, and responsibility for one's own thinking.

It allows disagreement without destruction. It protects the right to question, to dissent, and to exist without justification. These protections are not abstract ideals; they are practical safeguards against abuse of power.

A free society is not defined by uniformity, but by its ability to accommodate difference. People belong to many identities simultaneously—cultural, professional, spiritual, and personal. These identities overlap and coexist. Protecting one group's rights reinforces the principle that rights are inherent, not granted by favor or conformity.

When people understand this, manipulation loses its power. Division fails when individuals recognize shared interests beneath surface differences. Prejudice weakens when people refuse to accept borrowed hatred. Fear diminishes when complexity is acknowledged rather than denied.

Resisting manipulation does not require genius. It requires awareness. It requires asking simple but difficult questions: Who benefits from this belief? What evidence supports it? Who is being excluded, and why? What happens if this logic is applied further? These questions slow emotional reaction and restore agency.

Courage is essential, not the dramatic kind celebrated in stories, but the quiet courage of speaking when silence is easier, of listening when anger is tempting, and of standing with others when doing so carries risk.

This courage protects not only those who are targeted, but those who might be targeted next.

The temptation to surrender freedom for comfort will always exist. Times of uncertainty make it stronger. The quiet trap waits patiently, offering reassurance and certainty at the cost of vigilance. Escaping it requires deliberate effort and shared responsibility.

Freedom survives not because it is guaranteed, but because people choose it repeatedly. They choose to think rather than repeat, to **Question** rather than accept, to protect dignity rather than trade it. They choose solidarity over suspicion and truth over convenience.

Manipulation will never disappear. It is a constant feature of human societies. The real **Question** is whether people recognize it in time and refuse to participate. Doing so is not an act of idealism. It is an act of preservation—of dignity, of agency, and of the fragile, powerful freedom that allows human beings to live together without fear.

Questions & Answers

Question 1: Why are small concessions and temporary justifications more dangerous to freedom than direct force?
Answer: Direct force is visible and therefore more likely to provoke resistance. Freedom, however, erodes most effectively through habit and gradual distortion. Each small exclusion or "practical lie" appears reasonable in isolation, yet together they establish a dangerous logic in which rights become conditional rather than inherent. Once this principle is accepted, the boundary protecting everyone weakens permanently, because the same justification can later be applied to any group.

Question 2: How does the language of protection function as a psychological barrier to recognizing manipulation?
Answer: Manipulators rarely present themselves as oppressors; instead, they adopt the language of protection—defending security, tradition, identity, or moral purity. This framing appeals to people's craving for clarity during times of uncertainty and fear. Individuals feel empowered and safeguarded, while in reality they are being guided toward conformity and obedience. This hollow sense of strength depends on surrendering the right to dissent in exchange for belonging to something that feels "pure" and powerful.

Question 3: In what way does the irrelevance of truth serve manipulators better than a successfully proven lie?
Answer: Modern manipulation does not always aim to prove a false claim correct; its deeper goal is to undermine trust in truth itself. By discrediting institutions of inquiry—such as education, expertise, and independent thought—manipulators make evidence seem arrogant and questioning appear disloyal. Through constant repetition rather than proof, truth becomes

ineffective even when presented, because people no longer recognize or trust the processes that reveal it.

Question 4: Why is silence considered a signal of permission rather than a neutral act of survival?
Answer: Although silence is often justified as self-protection or avoidance of conflict, it is never truly neutral. Silence teaches those who exploit others how far they can proceed without resistance. When people remain silent as the "first group" is targeted, they unintentionally delay confrontation while allowing injustice to grow stronger. By the time the threat reaches them, the social trust and unity required for effective resistance may already be broken.

Question 5: How does the labor-saving nature of prejudice prevent societies from solving structural problems?
Answer: Prejudice offers an attractive shortcut because it simplifies complex realities and removes the need for learning or self-reflection. Instead of examining economic systems, power structures, or leadership failures, people are encouraged to blame those who are different. This diverts collective energy into identity-based conflict, leaving no strength to **Question** or reform the systems and authorities that benefit most from division.

End of the Chapter

Chapter–4.
The Power of Words That Shape a Life

Words carry more weight than we often realize. This chapter reveals how language influences thoughts, relationships, and identity itself. By understanding the power behind what you say, you'll learn to communicate with intention, build trust, and create lasting impact through clarity, honesty, and mindful expression.

There comes a moment in nearly every life when a person realizes that what they think and what they can say are not the same thing. The thought may feel complete, even powerful, inside the mind, yet when the time comes to express it, the words refuse to cooperate. Sentences trail off.

Meanings blur. What felt certain becomes fragile the moment it is spoken aloud. Many people walk away from such moments believing they lack confidence or intelligence, or that they are simply not "good with words."

What they rarely understand is that they have encountered a gap not in thought, but in vocabulary. Words are not ornaments added after thinking is done. They are the very framework that allows thinking to take shape and move outward into the world.

Language is the primary instrument through which human beings convert inner experience into shared reality. Every ambition, every fear, every plan, every protest, and every dream must pass through words to reach another mind. When vocabulary is limited, the range of expression is limited as well. Ideas that could have persuaded remain unheard. Feelings that could have been understood remain misinterpreted. Opportunities that required clarity quietly pass by. Vocabulary does not simply influence communication; it determines what communication is possible.

The mind works through symbols, and words are its most precise symbols. Just as a builder cannot create a complex structure with only a hammer, a thinker cannot develop complex ideas with only a handful of words. Precision matters. A vague word produces a vague thought. An exact word sharpens perception. When people acquire new words, they do not merely add labels to old ideas; they gain access to new distinctions. They begin to notice differences that once felt indistinguishable. They learn to separate assumption from evidence, feeling from fact, intention from impact. In this way, vocabulary expands not only speech, but awareness itself.

Words also shape judgment. When a person lacks the language to describe a situation accurately, they often rely on emotional shortcuts. Frustration becomes anger. Confusion becomes blame. Discomfort becomes fear. A richer vocabulary allows more nuanced interpretation. A person who can name uncertainty, ambiguity, or complexity is less likely to react impulsively. Language creates space between stimulus and response. In that space, judgment grows.

The adaptability of language is one of its greatest strengths. Human beings instinctively adjust their speech depending on audience and situation. The tone used with a close friend differs from the tone used in a formal meeting, not because one is dishonest, but because communication is contextual. Effective language respects circumstance. It considers who is listening, what they know, and what they need to understand. A developed vocabulary provides the flexibility required to move comfortably between these contexts. It allows a person to be clear without being crude, formal without being distant, persuasive without being manipulative.

Vocabulary also governs credibility. People instinctively judge the reliability of information by how it is expressed. Clarity suggests preparation. Precision suggests understanding. Hesitation and vagueness, even when ideas are sound, create doubt. This does not mean that complex words are superior to simple ones. It means that the right words—whether simple or sophisticated—signal mastery of thought. A strong vocabulary gives the speaker or writer command over emphasis, rhythm, and nuance, all of which influence how ideas are received.

One of the most overlooked truths about vocabulary is its influence on self-respect. When people repeatedly fail to express themselves accurately, they begin to doubt their own thinking. They grow silent in discussions, not because they have nothing to contribute, but because they fear being misunderstood. Over time, this silence can harden into disengagement. Conversely, when people acquire the words they need, confidence grows naturally. Not the loud confidence of domination, but the quiet confidence of being able to stand behind one's own thoughts.

Vocabulary development is inseparable from attention. Words do not announce themselves as important. They appear in passing—during conversation, in reading, in listening. Most people grasp approximate meaning and move on. Those who build vocabulary do something different. They pause. They notice. They ask why this word was chosen instead of another. They recognize that precision is intentional. This habit of attention turns ordinary exposure into education.

Context is the teacher that makes words usable. A definition alone rarely fixes a word in memory.

Meaning becomes clear when a word is seen performing its function inside a sentence, interacting with other ideas. Context reveals tone, implication, and limitation. It teaches what a word can do and what it cannot. Learning vocabulary through context also prevents misuse, which is often the source of embarrassment and hesitation.

Synonyms reveal the subtlety of language. Many words appear interchangeable at first glance, yet each carries a distinct shade of meaning. One may imply judgment, another neutrality. One may suggest intensity, another restraint. Choosing between them is not trivial; it shapes the emotional color of a message. This is why effective communication depends not only on knowing words, but on understanding their weight. Vocabulary depth—the ability to sense these differences—is what separates adequate expression from compelling expression.

Deliberate practice transforms vocabulary from passive recognition to active command. Writing down new words, reflecting on their meaning, and using them correctly in speech or writing embeds them into working memory. This process strengthens mental discipline. It trains the mind to value accuracy and resist approximation. Over time, expression becomes more deliberate and less reactive. Words are chosen rather than stumbled upon.

How to develop vocabulary for better communication and understanding?

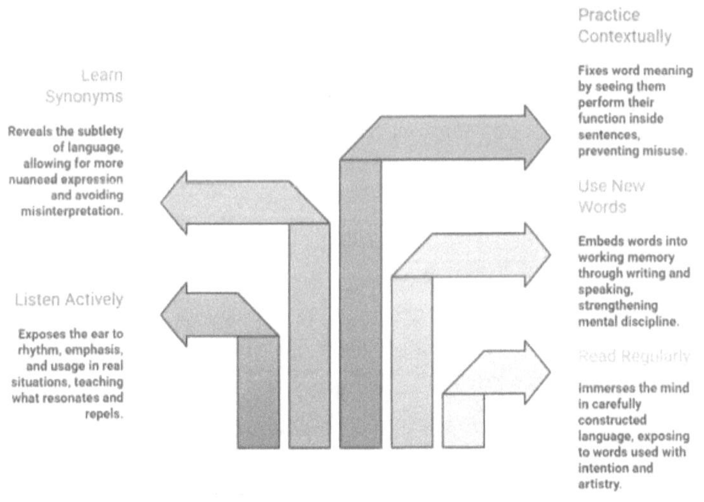

Vocabulary is also a gateway to participation. Every community, profession, and discipline develops its own language. Without familiarity with that language, individuals remain outsiders, regardless of intelligence or interest. Understanding terminology allows entry into conversation. It allows questions to be asked in ways that invite serious answers. It allows disagreement without confusion. In this sense, vocabulary is not merely educational; it is social.

The impact of vocabulary on relationships is profound. Misunderstandings often arise not from disagreement, but from imprecision. People argue past one another because they use the same words to mean different things, or different words to mean the same thing. A richer vocabulary reduces these collisions. It allows people to clarify intent before conflict escalates. It fosters empathy by enabling accurate expression of feelings and needs.

Writing exposes the full importance of vocabulary because writing offers no immediate correction. The writer must anticipate misunderstanding and prevent it through clarity. A limited vocabulary leads to repetition and generalization. A developed vocabulary allows structure, emphasis, and flow. Writing becomes not merely a task, but a craft. Ideas unfold logically because language supports them.

Education thrives or fails on vocabulary. Students often struggle with subjects not because concepts are beyond them, but because the language of those concepts is unfamiliar. When vocabulary improves, comprehension accelerates. Instructions become intelligible. Problems become solvable. Learning becomes cumulative rather than fragmented. Vocabulary is the foundation upon which all other learning rests.

Professional environments reward clarity relentlessly. Meetings, proposals, negotiations, and evaluations depend on language.

Those who express themselves precisely save time, reduce error, and inspire confidence. This is not manipulation; it is competence. Employers and colleagues respond to those who can translate ideas into action through words.

Vocabulary also enriches inner life. Emotions that remain unnamed often feel overwhelming. When people learn to distinguish between irritation and resentment, anxiety and fear, disappointment and despair, emotional control improves. Language allows people to understand themselves, not just others. In this way, vocabulary supports mental health as well as communication.

The growth of vocabulary is slow, but its effects are lasting. There is no moment when it is "finished." Language evolves, and so must those who use it.

New ideas require new words. Changing realities demand updated expression. Staying alert to language keeps thought flexible and relevant.

Importantly, vocabulary should never become a weapon of exclusion. Words are meant to connect, not to intimidate. The goal is understanding, not superiority. True mastery of language includes the ability to simplify without losing meaning. It includes the humility to choose clarity over display.

A person with a strong vocabulary can navigate complexity without confusion and simplicity without condescension. They can explain difficult ideas plainly and simple ideas beautifully. This adaptability is the mark of genuine intelligence.

Listening is as essential as speaking in vocabulary development. Listening exposes the ear to rhythm, emphasis, and usage. It teaches how language functions in real situations. It reveals what resonates and what repels. Listening well is a form of respect, but it is also a form of study.

Reading remains one of the most powerful tools for vocabulary growth because it immerses the mind in carefully constructed language. Through reading, one encounters words used with intention, precision, and artistry. Exposure to varied styles and subjects broadens both language and thought.

Vocabulary is power, but not power over others. It is power to participate fully in life.

It allows individuals to defend ideas, challenge injustice, explain themselves, and understand the world with clarity. It transforms silent thought into shared meaning.

To build vocabulary is to invest in oneself. It strengthens thinking, sharpens judgment, and amplifies voice. It opens doors that remain closed to those who cannot articulate their ideas, not because they lack worth, but because they lack language.

Words shape reality because they shape understanding. Those who choose to develop their vocabulary choose to live with greater precision, confidence, and connection. As words expand, so does the world they can engage with, influence, and understand.

Questions & Answers

Question 1: Why does the lack of a specific word often lead to emotional shortcuts such as anger or blame?
Answer: When a person cannot accurately describe a complex inner or outer situation, the mind defaults to broad emotional reactions to manage discomfort. Without words to name experiences like uncertainty, ambiguity, or tension, the brain interprets this vagueness as a threat and responds impulsively. A richer vocabulary provides precise labels that convert a vague emotional surge into a clear, manageable fact, reducing internal escalation and improving self-control.

Question 2: How does the habit of attention transform passive listening into a tool for cognitive growth?
Answer: Most people process language passively, absorbing only approximate meanings. In contrast, those who actively build vocabulary practice intentional attention: they notice why a particular word was chosen, how it differs from its alternatives, and what emotional or intellectual weight it carries. This habit deepens understanding, allowing language to move from simple recognition to strategic, conscious use, thereby strengthening thinking itself.

Question 3: In what way is the ability to simplify a concept a sign of superior vocabulary mastery?
Answer: True command of language is not demonstrated by complexity for its own sake, but by clarity. A person with a strong vocabulary can move fluidly between contexts, choosing simple words without losing depth or accuracy. This flexibility reflects genuine intelligence—the capacity to translate difficult ideas into plain language—ensuring that vocabulary serves as a bridge for understanding rather than a barrier.

Question 4: Why is a lack of vocabulary often mistaken for a lack of confidence or intelligence?

Answer: When ideas feel clear internally but become blurred or incomplete when spoken, individuals often assume the problem lies in their intelligence or courage. In reality, the difficulty is frequently structural: without precise words, the mind cannot construct a complete, shareable idea. Repeated failures to be understood may lead people to withdraw into silence, mistakenly believing they have nothing valuable to contribute.

Question 5: How does vocabulary function as a foundation for learning subjects beyond language itself?
Answer: Many academic struggles arise not from the complexity of concepts, but from unfamiliarity with the language used to explain them. Every discipline relies on its own vocabulary, and without understanding those terms, instructions remain unclear and problems unsolvable. Strengthening vocabulary therefore accelerates learning across all fields, making complex subjects more accessible and comprehension more efficient.

End of the Chapter

Chapter–5.
Developing Judgment, Responsibility, and Direction

A purposeful life is built on clear thinking and accountability. This chapter focuses on strengthening judgment, embracing responsibility, and choosing direction with intention. It shows how disciplined decisions and ownership of actions lead to confidence, stability, and a stronger sense of control over your future.

From the moment a human being enters the world, life begins as an experience of complete dependence. Every need is met by others, every danger is removed by someone else's hands, every comfort arrives from an external source. This early dependence is not weakness; it is the natural starting point of human existence. Yet life does not remain in that state. Growth, in its truest sense, is the gradual and often demanding movement away from dependence toward personal capability. Hidden within this movement is one of the most powerful forces shaping a successful and meaningful life: self-reliance.

Self-reliance is not the rejection of others, nor is it the denial of community or support. It is the inner readiness to stand responsibly within life rather than leaning passively against it. It is the ability to meet challenges with one's own effort, to think independently, to decide consciously, and to act with integrity even when guidance is absent. A self-reliant person does not wait for life to arrange itself conveniently; instead, they engage with it directly, shaping their own response to circumstances rather than surrendering control.

Many people imagine self-reliance as something dramatic, as if it appears suddenly through a bold act or decisive moment. In reality, it grows quietly. It is formed in ordinary situations where a person chooses effort instead of ease, responsibility instead of excuse, reflection instead of imitation. It is built through repetition. Small acts of independence, repeated consistently, gradually become habits. Habits become character, and character becomes destiny.

One of the earliest obstacles to self-reliance is comfort. Comfort can be a gift, but it can also become a subtle trap. When tasks are constantly done for someone, when decisions are always made on their behalf, the opportunity to grow is quietly removed. What seems like care can unintentionally teach helplessness. Over time, the individual learns that effort is optional, that responsibility can be delayed, and that someone else will step in if things become difficult. This lesson, learned early, can follow a person into adulthood and quietly limit their potential.

Self-reliance begins with the acceptance of responsibility. Responsibility is often misunderstood as punishment or pressure, but in truth it is empowerment. To take responsibility is to claim ownership of one's life. It means recognizing that choices have consequences and that those consequences belong to the person who makes them. When responsibility is accepted, excuses lose their power. Life no longer feels like something that happens to a person; it becomes something they actively participate in.

Taking responsibility changes how a person views time, effort, and commitment. Time becomes something to manage rather than waste. Effort becomes an investment rather than a burden. Commitments become promises to oneself, not obligations imposed by others. Through this shift, even simple tasks gain meaning. Completing work without reminders, preparing ahead of time, and honoring one's word are no longer external demands; they are expressions of self-respect.

As responsibility becomes a habit, confidence begins to grow naturally. This confidence is not loud or boastful. It does not depend on approval or praise.

It comes from knowing, through experience, that one can be trusted to act. Each responsibility met becomes evidence of capability. Over time, this evidence forms a stable sense of self-worth, grounded not in opinion but in proven effort.

Alongside responsibility, self-reliance requires the discipline of being informed. Independence without understanding is not strength; it is risk. To be self-reliant, a person must actively seek knowledge. This does not mean knowing everything, but it does mean refusing to remain ignorant out of convenience. An informed individual understands that decisions carry weight, and that good decisions require clarity.

Being informed involves curiosity and attentiveness. It means asking questions rather than accepting assumptions. It means listening carefully, observing thoughtfully, and reflecting honestly. Learning does not end with formal education; it continues through every conversation, experience, and challenge. A self-reliant person treats life itself as a teacher, remaining open to insight while remaining critical of misinformation.

Information strengthens judgment. When a person understands their circumstances, options, and potential outcomes, fear diminishes.

Uncertainty becomes manageable. Knowledge replaces guesswork, and clarity replaces confusion. This does not guarantee perfect decisions, but it allows decisions to be made with intention rather than impulse.

Knowing where one is going is another essential element of self-reliance. Direction gives purpose to effort. Without direction, even the hardest work can feel meaningless.

Direction does not require a flawless plan or complete certainty. It begins with reflection. It asks a person to consider their values, interests, and long-term hopes. It requires honesty about what truly matters, rather than what simply appears impressive or socially acceptable.

Many people allow their direction to be shaped entirely by others. Friends, family, trends, and expectations often influence choices. Guidance can be valuable, but surrendering direction entirely weakens personal identity. A self-reliant person listens carefully to advice but does not abandon judgment. They understand that no one else can fully live their life or bear the consequences of their choices.

Discovering direction is often uncomfortable. It involves uncertainty and doubt. It may require admitting confusion or changing course. Yet this discomfort is not failure; it is growth. Each thoughtful decision, even when imperfect, sharpens self-awareness. Over time, direction becomes clearer not because life becomes simpler, but because the individual becomes more grounded.

Components of Self-Reliance

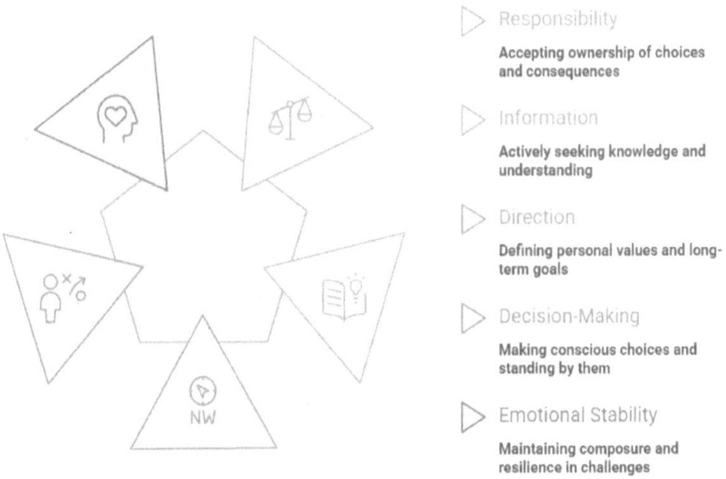

Decision-making is where self-reliance becomes visible. Making decisions requires courage. It involves accepting the possibility of error and standing behind outcomes rather than hiding behind excuses. Many people avoid decisions because they fear being wrong. Yet refusing to decide is itself a decision, one that often leads to stagnation.

Mistakes are not enemies of self-reliance; they are its teachers. A person who never decides for themselves never learns from consequences. Growth requires the willingness to choose, reflect, adjust, and choose again. Each decision refines judgment and builds resilience. Over time, confidence becomes steadier, grounded not in perfection but in experience.

Self-reliant decision-making does not reject collaboration or guidance. It simply refuses dependence. A self-reliant person may seek input, consider perspectives, and weigh options, but the final responsibility remains their own. This balance preserves humility without sacrificing independence. It allows learning without surrendering agency.

As self-reliance deepens, it reshapes behavior in subtle but powerful ways. A self-reliant individual does not wait to be told what needs to be done.

They notice problems and consider solutions. They do not complain endlessly about obstacles; they look for alternatives. Initiative replaces passivity. Action replaces frustration. Life becomes something to engage with rather than endure.

In social situations, self-reliance brings emotional stability. A self-reliant person is less easily controlled by approval or criticism. Their sense of worth does not rise and fall with opinion. This does not mean they ignore feedback; it means they evaluate it thoughtfully. They know who they are becoming, and they measure themselves against that internal standard rather than against others.

In work and study, self-reliance encourages excellence. A self-reliant person prepares thoroughly, manages responsibilities independently, and seeks improvement continuously. Motivation comes from within rather than from external pressure. This quality makes them dependable and adaptable. Trust grows naturally around them, not because they demand recognition, but because their actions consistently demonstrate reliability.

Emotionally, self-reliance fosters resilience. Life inevitably brings disappointment, uncertainty, and hardship. A self-reliant person does not collapse at the first sign of difficulty. They acknowledge emotions without being ruled by them. They trust their ability to endure, adapt, and recover. This emotional steadiness does not remove pain, but it prevents despair from taking root.

It is important to understand that self-reliance does not mean isolation. Human beings thrive through connection. We learn from one another, support one another, and grow together. Self-reliance simply ensures that relationships are built on strength rather than neediness. When individuals rely on themselves, they contribute more fully to others. They offer support rather than demand rescue, cooperation rather than dependency.

The habit of self-reliance is not developed through a single decision, but through daily practice. Each day presents opportunities to either strengthen or weaken it. Choosing to manage responsibilities independently, to seek understanding rather than shortcuts, to reflect before deciding, and to act with intention reinforces self-reliance. Over time, these choices become natural. What once required effort becomes character.

There will always be moments when seeking help is wise and necessary. Asking for help is not a failure of self-reliance; refusing to grow is. The difference lies in intention. A self-reliant person seeks help to learn and improve, not to avoid responsibility. They accept guidance while maintaining ownership of their path.

As self-reliance matures, it brings a quiet satisfaction. Life feels more coherent and purposeful. Decisions align with values. Effort feels meaningful. Challenges become opportunities for growth rather than threats to comfort. The individual experiences a sense of authorship over their own life, understanding that while not everything can be controlled, much can be shaped through thought and action.

The true reward of self-reliance is dignity. It is the dignity of standing on one's own effort, of facing life honestly, and of growing through experience. A self-reliant person understands that success is not measured by ease or convenience, but by engagement and contribution. They do not merely get by; they build, learn, and evolve.

Developing self-reliance is a lifelong journey. It does not end with youth or education. Each new stage of life brings new responsibilities and challenges, inviting deeper reliance on judgment, discipline, and character. Those who accept this invitation grow steadily, not because life becomes simpler, but because they become stronger.

In the end, self-reliance is an act of respect—respect for one's own potential. It is the decision to live deliberately, to learn continuously, and to take responsibility for the direction of one's life. It is demanding and often uncomfortable, but profoundly rewarding. It shapes not only what a person achieves, but who they become.

Questions & Answers

Question 1: Why is self-reliance presented as a gradual habit rather than a single decisive act?
Answer: The chapter emphasizes that self-reliance is not formed through dramatic moments but through repeated, ordinary choices. Growth happens when individuals consistently choose effort over ease, responsibility over excuses, and reflection over imitation. These small decisions accumulate into habits, habits shape character, and character ultimately determines destiny. By framing self-reliance as gradual, the text highlights that true independence is built quietly through discipline, not achieved suddenly through bold gestures.

Question 2: How can comfort, though well-intentioned, become an obstacle to personal growth?
Answer: Comfort becomes harmful when it removes opportunities for effort and decision-making. When others constantly solve problems or make choices on someone's behalf, the individual is denied the chance to develop competence and judgment. What appears as care can unintentionally teach helplessness, reinforcing the belief that responsibility can be avoided and effort is unnecessary. Over time, this learned dependence weakens initiative and limits potential well into adulthood.

Question 3: Why does the chapter redefine responsibility as empowerment rather than burden?
Answer: Responsibility is described not as pressure imposed from outside, but as ownership claimed from within. Accepting

responsibility allows individuals to see themselves as active participants in life rather than passive recipients of circumstances. This shift removes the power of excuses, transforms effort into investment, and turns commitments into expressions of self-respect. Responsibility empowers because it restores control, giving meaning to action and dignity to effort.

Question 4: What role does being informed play in strengthening judgment and reducing fear?
Answer: Self-reliance without understanding is portrayed as dangerous rather than strong. Being informed equips individuals to make decisions based on clarity instead of impulse or guesswork. Knowledge reduces fear by making uncertainty manageable and transforming confusion into choice. While information does not guarantee perfect outcomes, it enables intentional decision-making and strengthens judgment by replacing emotional reaction with thoughtful evaluation.

Question 5: How does self-reliance balance independence with connection rather than leading to isolation?
Answer: The chapter clearly distinguishes self-reliance from isolation. Human growth depends on relationships, learning, and cooperation, but self-reliance ensures these connections are based on strength instead of neediness. A self-reliant person seeks help to grow, not to escape responsibility. This balance allows individuals to contribute meaningfully to others, offering support rather than demanding rescue, and building relationships grounded in mutual respect rather than dependency.

End of the Chapter

Chapter-6.
Manners as a Measure of Character and Respect

True character often reveals itself in the smallest gestures. This chapter highlights how manners go beyond etiquette, reflecting respect, awareness, and inner values. It demonstrates how simple acts of courtesy can shape relationships, influence perception, and quietly define the kind of person you are becoming.

Cities rise with steel and stone, but they endure through something far less visible and far more powerful: the human ability to live together. Long before roads were paved and buildings reached the sky, people discovered that survival and happiness depended not only on strength or skill, but on cooperation. Living together required understanding, patience, and a shared respect for one another.

Out of this need grew customs that helped people coexist peacefully. These customs, refined over generations, are what we call manners. Though often dismissed as trivial or old-fashioned, manners remain one of the most essential forces holding human relationships together.

Manners are not artificial rules imposed to control behavior; they are natural expressions of consideration. Wherever people genuinely care about one another, manners appear almost instinctively. They exist in the way a neighbor respects another's space, in the patience shown while waiting one's turn, and in the effort to make others feel welcome and at ease. Often, people practice good manners without ever labeling them as such. They simply act with awareness that their behavior affects others. In this sense, manners are not about formality but about thoughtfulness.

From early childhood, manners quietly shape social life. A child who learns to share, to listen, and to apologize is not merely learning rules; they are learning how relationships work. These early lessons teach that one's actions carry consequences beyond oneself. When a child is considerate, trust grows. When they are careless, tension follows. Over time, these experiences form an understanding that cooperation makes life smoother, warmer, and more enjoyable for everyone involved.

Yet manners are often misunderstood, especially by the young. They can appear restrictive, unnecessary, or even unfair. Children may see them as tools used by adults to control behavior rather than as skills that protect harmony. This misunderstanding is not surprising. The value of manners is rarely felt when everything is going well. It becomes visible only when manners are absent. A slammed door, an ignored greeting, or a careless interruption may seem small, but their effects ripple outward, creating discomfort, resentment, and distance.

When manners are neglected, everyday life becomes strained. Small acts of inconsideration accumulate, slowly eroding goodwill. A person who constantly interrupts teaches others that their voices do not matter. Someone who ignores shared rules suggests that their convenience is more important than fairness. Over time, such behavior isolates the individual, even if they never intend to be rude. People withdraw not out of anger, but out of fatigue. Few wish to remain close to someone who consistently makes them uncomfortable.

Manners are especially important in shared spaces. Homes, schools, workplaces, and communities function smoothly only when individuals recognize their mutual dependence. No one exists entirely alone. Each person's comfort is linked to the behavior of others. Cleanliness, order, and courtesy are not merely personal choices; they are shared responsibilities. When one person neglects them, everyone feels the impact. Manners serve as the silent agreements that protect shared life from chaos.

At the heart of manners lies consideration. Consideration is the ability to pause and think beyond oneself. It asks simple but powerful questions: How will my action affect others? Is this fair? Is this kind? These questions do not require extraordinary effort, only awareness. Yet they transform behavior. A considerate person naturally waits their turn, speaks respectfully, and acknowledges others' presence. They do not act out of fear of punishment, but out of understanding.

Table manners provide a clear example of how consideration improves daily life. Eating together is one of the oldest human rituals. When people approach it carelessly, meals become uncomfortable and tense. When they approach it with courtesy, meals become moments of connection. Cleanliness, patience, and pleasant conversation are not rigid rules; they are gestures that show respect for those sharing the table. Through such simple acts, manners turn routine moments into meaningful ones.

Manners also play a crucial role in hospitality. Guests enter another person's space, and with that entry comes vulnerability. A polite welcome, attentive listening, and thoughtful behavior signal safety and respect.

When manners are ignored, guests feel unwelcome and uneasy. Hospitality, at its core, is an expression of empathy. It recognizes that being welcomed with kindness strengthens bonds and builds trust.

For young people, manners are often tested in social settings where impulses are strong and awareness is still developing. A sudden desire to join a game, speak out, or assert oneself can override consideration. In such moments, the absence of manners often leads to conflict. Arguments arise, feelings are hurt, and enjoyment fades. These experiences, though uncomfortable, serve as powerful lessons. They reveal that personal satisfaction gained at the expense of others rarely lasts.

The realization that one's own discomfort can be caused by someone else's bad manners is often the turning point in understanding their importance. Being interrupted, excluded, or disrespected creates immediate awareness of how fragile social harmony can be. In such moments, manners cease to feel like abstract rules and become personal needs. A person begins to see that the same behavior they dismiss in themselves becomes painful when directed at them.

Apologies are among the most meaningful expressions of good manners. They require humility and courage. To apologize is to acknowledge that one's actions affected another person negatively. Many find this difficult because it challenges pride. Yet apologies repair relationships in ways few other actions can. They reopen communication, restore trust, and demonstrate growth. Like all manners, apologizing improves with practice. The more it is used sincerely, the more natural it becomes.

Manners are not limited to politeness; they reflect character. A person's behavior in moments of inconvenience or frustration reveals their true regard for others. Courtesy shown only when it is easy is fragile. True manners endure under pressure. They guide behavior when emotions are high and patience is low. In this way, manners become a form of self-discipline, helping individuals act thoughtfully rather than react impulsively.

In modern life, where speed and efficiency are often prioritized, manners are sometimes viewed as outdated. Yet the pace of life makes them more necessary, not less. Crowded cities, digital communication, and constant interaction increase the chances of misunderstanding and offense. Manners act as stabilizers, reducing friction and preserving dignity. They remind people that behind every interaction is a human being deserving of respect.

Manners also teach fairness. Standing in line, taking turns, and respecting shared rules ensure that no one is unfairly disadvantaged. These practices reinforce the idea that everyone's time and presence matter equally. When such norms are ignored, resentment grows. Fairness, supported by manners, creates trust in systems and relationships alike.

Importantly, manners are not about suppressing individuality. They do not demand uniform behavior or silence honest expression. Instead, they provide a framework within which individuality can flourish without harming others. A person can be confident, outspoken, and assertive while remaining courteous. Manners guide how something is expressed, not whether it is expressed.

Children often learn manners not through instruction alone, but through observation. The behavior modeled by adults leaves lasting impressions. When children see respect practiced consistently, they absorb it naturally. When they see inconsideration excused or rewarded, they learn those patterns as well. Teaching manners, therefore, requires example as much as explanation.

As individuals mature, manners evolve. What begins as simple politeness grows into deeper social awareness. Adults learn to navigate complex relationships, cultural differences, and emotional sensitivities.

Manners adapt to context, requiring flexibility and empathy. In this way, manners remain relevant throughout life, shaping interactions at every stage.

Ultimately, manners are a form of social intelligence. They reflect an understanding of human connection and the willingness to protect it. They reduce conflict, build trust, and create environments where cooperation thrives.

A society without manners may function through force or rules alone, but it cannot flourish.

The **Question** of whether manners are important is, at its core, a **Question** about how we choose to live with one another. Manners represent the decision to acknowledge shared humanity. They affirm that personal freedom does not require disregard for others. Instead, true freedom is enhanced by mutual respect.

Manners do not demand perfection. Everyone falters. What matters is the willingness to learn, reflect, and improve. Each act of consideration strengthens the social fabric, making life more livable and relationships more rewarding.

In the end, manners are not about impressing others or following outdated customs. They are about creating a world in which people feel seen, respected, and valued. They are quiet, often unnoticed, but deeply powerful. And in every age, in every community, they remain essential to living well together.

Questions & Answers

Question 1: Why does the chapter describe manners as a foundation of social survival rather than a set of optional social rules?

Answer: The chapter frames manners as emerging from humanity's earliest need to live together peacefully. Long before formal laws or institutions, cooperation determined survival. Manners developed as practical tools to reduce conflict, protect dignity, and enable coexistence. By presenting manners as rooted in necessity rather than tradition, the chapter shows that they are not optional decorations of behavior, but essential mechanisms that allow communities, cities, and relationships to endure.

Question 2: How does the chapter explain the difference between manners as control and manners as consideration?

Answer: Manners are often misunderstood as external rules used to restrain behavior, especially by children. The chapter challenges this view by defining manners as expressions of consideration—an internal awareness that one's actions affect others. True manners arise not from fear of punishment but from understanding. This distinction shifts manners from being imposed limits to being voluntary acts of respect, transforming them into reflections of character rather than obedience.

Question 3: Why do small acts of discourtesy have a disproportionate impact on relationships and social harmony?

Answer: The chapter emphasizes that minor acts—interruptions, ignored greetings, careless behavior—accumulate over time. While each act may seem insignificant, together they erode goodwill and trust. People rarely withdraw from relationships because of one major offense; instead, they distance themselves due to repeated discomfort. This explains why manners function as preventive forces, preserving harmony by addressing small frictions before they grow into lasting resentment.

Question 4: In what way do apologies reveal the deeper moral dimension of manners?

Answer: Apologies are highlighted as one of the most meaningful expressions of manners because they require humility, self-awareness, and courage. To apologize is to acknowledge responsibility for harm caused to another. This act repairs trust and restores communication, demonstrating that manners are not superficial politeness but moral actions rooted in accountability. The difficulty of apologizing reveals how closely manners are tied to pride, character, and emotional maturity.

Question 5: How does the chapter reconcile manners with individuality and personal freedom?

Answer: The chapter rejects the idea that manners suppress individuality. Instead, it argues that manners provide a framework that allows individuality to exist without harming others. Confidence, assertiveness, and honesty are not incompatible with courtesy; manners guide *how* expression occurs, not whether it occurs. This balance shows that true freedom is not the absence of restraint, but the presence of respect, enabling personal expression within a shared social space.

End of the Chapter

Chapter-7.
The Art of Meaningful Conversation

Conversation is more than exchanging words—it's a gateway to understanding and connection. This chapter explores how to engage thoughtfully, listen actively, and express ideas with clarity. By mastering meaningful dialogue, you can build trust, influence perspectives, and create relationships that are both deep and lasting.

Conversation is so deeply woven into daily life that its importance is often overlooked. People speak constantly—at home, at work, in public spaces, and now across digital platforms—yet genuine conversation remains rare. True conversation is not merely the exchange of words; it is the exchange of attention, understanding, and respect. It is the process by which individuals recognize one another as thinking, feeling human beings. Long before written language existed, conversation was the primary way knowledge, values, and culture were preserved. Even now, when information is abundant and instantly accessible, conversation remains the most powerful tool for human connection and personal growth.

At its heart, meaningful conversation begins with presence. Presence means more than physical proximity; it means mental and emotional engagement. A person may be standing nearby yet be absent in thought, distracted by internal concerns or external interruptions. Conversation loses its vitality when presence is missing. When people are truly present, however, conversation becomes alive. Words carry weight, pauses gain meaning, and even silence feels comfortable rather than awkward. Presence communicates respect before a single word is spoken.

Many assume that good conversation requires confidence, wit, or eloquence. In reality, it requires awareness. Awareness of oneself, awareness of others, and awareness of the moment. A meaningful conversation does not demand flawless speech or clever remarks. It asks only that participants care enough to listen, respond thoughtfully, and remain open to the exchange. Conversation thrives when it is treated as a shared experience rather than a performance.

A powerful conversation often feels effortless, yet it rests on invisible skills. One of the most important is courtesy. Courtesy is the quiet discipline of treating others with dignity. It shapes tone, timing, and choice of words. A courteous speaker considers how their words will land, not just what they wish to express. This does not mean avoiding honesty or disagreement; it means expressing truth without unnecessary harm. Courtesy transforms conversation from confrontation into cooperation.

Courtesy creates emotional safety. When people feel safe, they speak more openly. They take intellectual risks, admit uncertainty, and explore ideas more deeply. Without courtesy, conversation becomes guarded. People speak defensively or withdraw entirely. Even strong ideas lose their power when delivered without consideration. Courtesy does not weaken communication; it strengthens it by allowing ideas to be heard rather than resisted.

Closely connected to courtesy is the practice of inclusion. A conversation reaches its fullest potential when all participants feel they belong. Inclusion does not require equal speaking time, but it does require equal respect. When conversation becomes dominated by a single voice or limited to a select few, its richness fades. Diverse perspectives add depth, challenge assumptions, and prevent stagnation. Inclusion ensures that conversation remains dynamic and balanced.

Inclusion requires attentiveness to social cues. Some people speak readily, while others hesitate. A thoughtful conversationalist notices this difference. They do not force participation, but they create openings. A simple invitation, a pause, or a direct acknowledgment can encourage quieter voices to enter the exchange. When people feel invited rather than pressured, they contribute more authentically.

Respect for differing opinions is another cornerstone of meaningful conversation. Disagreement is not a threat to conversation; it is an opportunity. Through disagreement, ideas are tested and refined. However, disagreement must be handled with care. When opinions are dismissed or ridiculed, conversation collapses into conflict. Respect allows disagreement to exist without damaging relationships. It recognizes that intelligence and sincerity can exist alongside differing views.

Respect is demonstrated through listening. Listening is not merely waiting for one's turn to speak; it is the active effort to understand. A good listener gives full attention, observes emotional cues, and responds to what was actually said rather than what they expected to hear. This level of listening requires patience and humility. It asks a person to set aside the need to be right and instead seek clarity.

Listening is often underestimated because it is quiet. Yet its impact is profound. When someone feels truly heard, tension eases. Defensiveness fades.

Even difficult topics become manageable. Many conflicts persist not because of irreconcilable differences, but because one or both sides feel unheard. Listening restores balance by affirming the other person's experience.

Contribution is another essential element of good conversation. Conversation is a shared creation, not a solo act. Each participant brings unique experiences, insights, and questions. Contribution does not require brilliance or originality. It requires sincerity. Even a simple observation or honest **Question** can move conversation forward. Withholding contribution out of fear limits both personal growth and collective understanding.

At the same time, contribution must be balanced with restraint. Over-speaking can be as damaging as silence. When one person consistently redirects conversation toward themselves, others disengage. Balanced contribution involves responding to what has been said and building upon it rather than replacing it. This responsiveness keeps conversation cohesive and meaningful.

Staying focused is another often-overlooked skill. Conversations lose depth when they wander aimlessly. While some flexibility is natural, constant topic changes disrupt flow. Depth requires continuity.

When people remain with a subject long enough to explore it, conversation becomes more satisfying. Ideas evolve, connections emerge, and understanding deepens. Focus turns conversation from scattered remarks into a shared journey.

Emotional awareness plays a vital role in this journey. Every conversation carries emotional context. People bring moods, expectations, and experiences into dialogue. A sensitive conversationalist recognizes emotional shifts and adjusts accordingly. They know when to pursue a topic and when to pause. Timing, in conversation, is as important as content. A well-timed silence can be more meaningful than many words.

Poor conversation habits often develop unconsciously. Interrupting, dominating discussion, dismissing others, or changing topics abruptly may not be intentional. They often arise from impatience, insecurity, or lack of awareness. Yet their effects are real. These habits create frustration and distance. Recognizing them is the first step toward improvement. Growth in conversation begins with honest self-reflection.

The consequences of poor conversation are often subtle. People may feel overlooked or undervalued without understanding why. Social interactions may feel strained or unfulfilling. Opportunities for connection quietly disappear. Improving conversational habits can dramatically change these experiences, often without altering personality. Awareness alone can transform interaction.

Conversation is closely tied to self-reliance. A self-reliant individual takes responsibility for how they communicate. They do not blame misunderstandings solely on others. They examine their own tone, assumptions, and listening habits. They recognize that effective communication is a skill to be practiced, not a trait to be assumed. Through this responsibility, conversation becomes an avenue for personal growth.

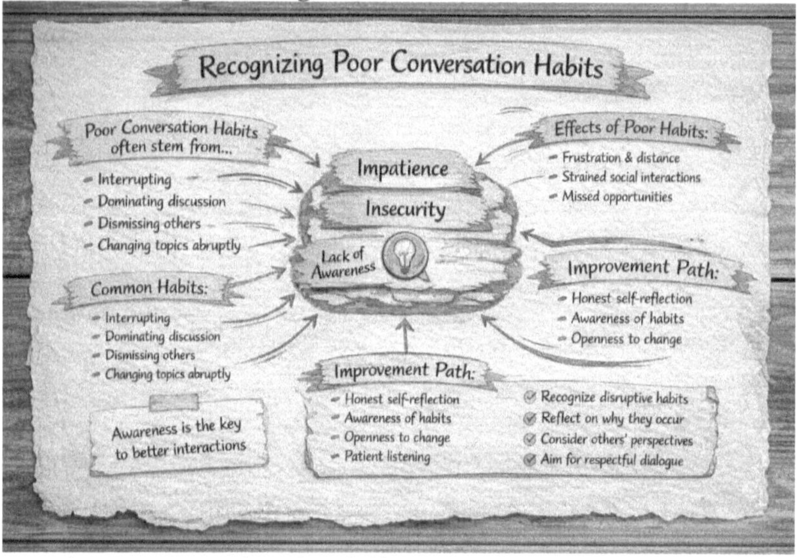

As conversational skill improves, confidence grows naturally. This confidence does not come from dominating discussion, but from trusting one's ability to engage meaningfully. A person who listens well, responds thoughtfully, and contributes sincerely feels at ease in conversation. Social anxiety diminishes when conversation is no longer a performance to manage but a connection to experience.

Conversation also serves as a powerful learning tool. Through dialogue, ideas are tested and refined. Exposure to different viewpoints expands understanding and challenges assumptions. Learning through conversation is dynamic; it engages both intellect and emotion. A meaningful conversation often lingers in the mind, prompting reflection long after it ends.

In group settings, effective conversation strengthens cooperation. Clear communication reduces misunderstanding and builds trust. Shared goals emerge naturally when ideas are openly discussed. When conversation is poor, confusion and resentment grow. When it is strong, collaboration feels effortless. This principle applies equally to families, classrooms, workplaces, and communities.

Conversation shapes identity. The way a person communicates influences how they are perceived and how they perceive themselves. A thoughtful conversationalist develops a reputation for empathy and reliability. Over time, this reputation opens doors to deeper relationships and greater influence. Conversation becomes not just interaction, but expression of character.

In modern life, where communication is rapid and often fragmented, meaningful conversation requires intention. Distractions compete for attention, and patience is tested. Yet the need for genuine connection has not diminished. If anything, it has intensified. People crave understanding amid noise. Conversation, when practiced with care, provides that understanding.

Improving conversation does not require perfection. Everyone missteps. What matters is willingness to learn and adapt. Each interaction offers an opportunity to practice presence, courtesy, inclusion, respect, focus, and listening. Over time, these practices become habits. Habits become strengths. Strengths shape relationships.

A good conversation leaves participants energized rather than drained. It creates connection rather than division. It feels spontaneous yet purposeful, relaxed yet attentive. When people leave a conversation feeling heard, respected, and enriched, something meaningful has occurred.

Ultimately, conversation reflects how individuals view one another. Careless conversation signals indifference. Thoughtful conversation signals respect. The choice is constant and personal. By cultivating better conversational habits, people improve not only their social lives, but also the quality of the communities they inhabit.

The art of meaningful conversation is not reserved for a select few. It is accessible to anyone willing to practice awareness and consideration. Through conversation, ideas are shared, relationships are built, and lives are shaped. In learning how to converse meaningfully, individuals learn not only how to speak and listen, but how to live more fully and responsibly with others.

Questions & Answers

Question 1: Why does the chapter distinguish meaningful conversation from the mere exchange of words?
Answer: The chapter emphasizes that conversation is not defined by speaking alone, but by the exchange of attention, understanding, and respect. Words without presence become noise rather than connection. Meaningful conversation requires mental and emotional engagement, where participants recognize one another as thinking and feeling individuals. This distinction highlights that true conversation is relational, not transactional, and that its value lies in human recognition rather than information transfer.

Question 2: How does presence function as the foundation of all meaningful conversation?
Answer: Presence is portrayed as the condition that gives conversation life. Without presence, even well-spoken words lose impact, and interaction becomes hollow. Presence involves attentiveness to the moment, the speaker, and the emotional context. When presence exists, pauses gain meaning, silence feels comfortable, and communication becomes fluid. The chapter suggests that presence itself communicates respect, establishing trust before any content is exchanged.

Question 3: Why are courtesy and respect described as strengthening conversation rather than limiting honesty or disagreement?
Answer: Courtesy and respect do not suppress truth; they shape how truth is delivered. The chapter argues that ideas, even strong or opposing ones, are more likely to be heard when expressed with consideration. Courtesy creates emotional safety, allowing participants to explore uncertainty and disagreement without defensiveness. In this way, respect transforms conversation from

confrontation into cooperation, enabling deeper intellectual and emotional exchange.

Question 4: In what way does active listening serve as the central discipline of meaningful conversation?
Answer: Listening is presented not as passivity, but as an active effort to understand. It requires attention, patience, and humility—the willingness to suspend one's need to be right. The chapter explains that many conflicts persist because people feel unheard rather than because they fundamentally disagree. Active listening reduces tension, affirms experience, and restores balance, making even difficult conversations manageable and productive.

Question 5: How does meaningful conversation contribute to personal growth and social cohesion?
Answer: The chapter links conversation to self-reliance, learning, and identity. By taking responsibility for how they communicate, individuals grow in awareness and confidence. Meaningful conversation becomes a tool for learning, refining ideas, and expanding perspective. On a social level, strong conversational habits build trust, reduce misunderstanding, and strengthen cooperation. Thus, conversation shapes not only relationships, but communities and character itself.

End of the Chapter

Chapter–8.
The Intelligent Use of Leisure Time

Leisure is not just rest—it's an opportunity for growth. This chapter teaches how to use free time wisely, balancing relaxation with activities that enrich the mind, nurture creativity, and strengthen personal skills. Intelligent leisure transforms idle hours into meaningful experiences that contribute to long-term success

Leisure time is one of the quietest forces shaping a person's life, yet it is also one of the most powerful. Unlike work or obligation, leisure arrives without instruction. No one tells us exactly how to use it, what to make of it, or how much it matters. And because it seems unimportant when compared with duties and responsibilities, it is often neglected, wasted, or treated as something to simply "get through." Yet how a person uses leisure time often reveals more about their direction, character, and future than how they perform under obligation. Work shows what we must do; leisure shows who we choose to become.

Many people believe they lack leisure time, but this belief rarely matches reality. Modern life, for all its pressures, has created more free time than any previous age. Advances in tools, technology, and organization have shortened work hours and reduced physical labor.

Tasks that once consumed entire days now take minutes. What has not changed, however, is the human challenge of deciding what to do with the time that remains. Leisure time, paradoxically, can feel like a burden when it lacks purpose.

When leisure time is left unplanned, it often dissolves into boredom. Boredom is not the absence of activity; it is the absence of meaning. A person may sit surrounded by options and still feel restless and dissatisfied. This restlessness is often misunderstood as bad luck or lack of opportunity, when in truth it is a signal. It signals unused potential and unclaimed freedom. Leisure time confronts us with choice, and choice requires effort.

Looking back through history offers useful perspective. Only a few generations ago, leisure was rare.

Long hours of labor filled most days, leaving little time or energy for personal interests. Free time was brief and precious, often used simply to rest and recover. Today, by contrast, leisure has expanded. Even those who feel busy often have hours each week that are unstructured. This expansion has created opportunity, but also responsibility. Time that once belonged to survival now belongs to choice.

The intelligent use of leisure time begins with a shift in attitude. Leisure is not empty time to be filled at random; it is living time. It belongs to the same life as work, learning, and responsibility. When treated carelessly, it weakens discipline and dulls ambition. When treated thoughtfully, it refreshes the mind, strengthens the body, and deepens understanding. Leisure time can either drain a person or develop them.

One of the most valuable functions of leisure is change. A healthy leisure activity provides a contrast to daily obligations. This change restores balance.

A person who spends the day engaged in mental work may find renewal in physical activity. Someone whose work is repetitive may seek creativity in their free time. This change is not escape; it is renewal. It prevents fatigue from becoming discouragement and routine from becoming stagnation.

Another essential value of leisure lies in learning. Leisure time offers freedom to explore interests without pressure. Learning during leisure is voluntary, and for that reason, it often becomes deeper and more satisfying than forced study. Skills developed during leisure—whether intellectual, creative, or practical—quietly build confidence. They expand competence and awaken curiosity. Over time, these skills often influence career paths, relationships, and self-respect.

Leisure also offers space for long-range growth. Not every leisure activity needs a grand purpose, but meaningful ones often point beyond the present moment. A person who practices a skill over time, who steadily improves, and who sets personal goals experiences a sense of progress. This progress gives leisure dignity. It transforms free time from distraction into investment. Even when the activity remains a hobby, the discipline it develops carries into other areas of life.

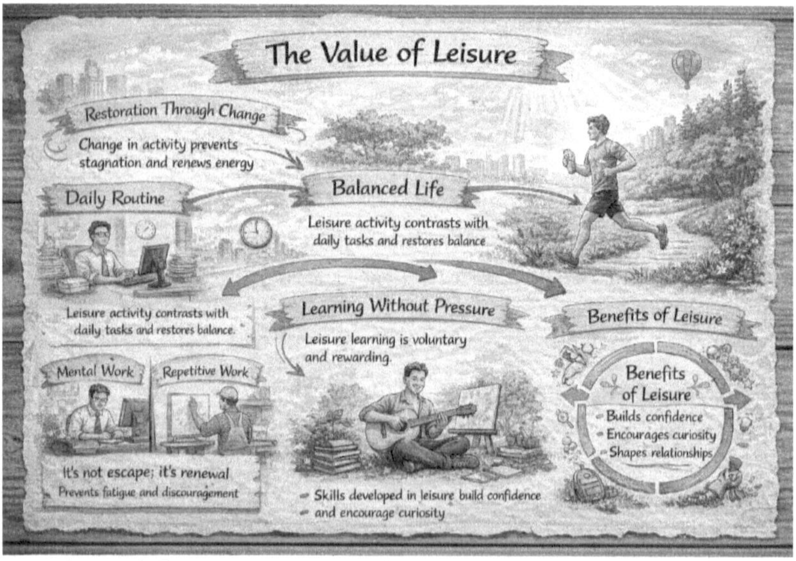

One of the common mistakes in leisure is waiting for others. When leisure depends entirely on companions, schedules, or external excitement, disappointment is frequent. Friends are busy. Plans fall through. Circumstances change. A person who has no independent interests becomes vulnerable to boredom and frustration. Self-directed leisure restores control. It allows enjoyment regardless of others' availability.

Hobbies illustrate this principle clearly. A hobby is not merely something to pass time; it is an activity chosen freely for satisfaction and growth. Whether creative, physical, intellectual, or practical, hobbies engage attention and effort. They invite patience and persistence.

Over time, they become sources of pride. The person who develops a hobby develops more than a skill; they develop a relationship with their own abilities.

Leisure activities also shape identity. What a person chooses to do when no one demands anything of them reflects their values. Some choices leave little behind beyond momentary pleasure. Others quietly shape habits, skills, and perspectives. Neither rest nor enjoyment is wrong; both are necessary.

But leisure becomes truly valuable when enjoyment and growth coexist.

Reading is one of the most powerful leisure activities available. It requires little equipment, adapts to any schedule, and opens countless worlds. Through reading, a person encounters ideas beyond their immediate experience. They learn not only facts, but ways of thinking. Reading during leisure time strengthens imagination, language, and judgment. It trains the mind to concentrate and reflect—skills increasingly rare and increasingly valuable.

Creative pursuits serve a different but equally important purpose. Drawing, writing, music, photography, and craftsmanship allow expression without external pressure. They help individuals translate thought and emotion into form. This process brings satisfaction not because of recognition, but because of creation itself. Creative leisure fosters patience and self-trust. It reminds people that they can make something meaningful with their own hands and minds.

Physical activities are another essential use of leisure. Movement refreshes both body and mind. Whether structured or informal, physical activity builds strength, coordination, and resilience. It relieves tension and improves mood.

Leisure that includes physical engagement supports health in ways no other activity can fully replace. It also teaches discipline and persistence, reinforcing habits that benefit every area of life.

Group activities offer social benefits when chosen thoughtfully. Teams, clubs, and shared projects create a sense of belonging and shared purpose. When participation is active rather than passive, group leisure strengthens cooperation and communication. It teaches individuals how to contribute, lead, and adapt. Such experiences are especially valuable when they balance enjoyment with responsibility.

Part-time work can also serve as meaningful leisure, particularly when chosen for learning rather than necessity alone. Work done voluntarily during free time teaches responsibility in a different context.

It builds experience, independence, and confidence. When approached thoughtfully, it prepares individuals for future roles while still respecting the spirit of leisure.

The key to better use of leisure time is not finding a perfect activity, but developing awareness. A person must ask simple but honest questions. Does this activity refresh me or drain me? Does it teach me something or leave me unchanged? Does it help me grow, even in small ways? These questions do not require strict answers, but they guide better choices.

Leisure does not need to be efficient, but it should be intentional. Intentional leisure recognizes that time is finite and valuable. It resists the habit of letting hours disappear without memory or meaning. This does not mean eliminating relaxation or amusement. It means choosing them consciously rather than by default.

Boredom, when examined closely, often signals a lack of direction rather than lack of opportunity. When a person feels bored, it is often because they are waiting to be entertained instead of engaging themselves. Learning to initiate activity is a skill, and like all skills, it improves with practice. The more a person uses leisure well, the easier it becomes.

Parents and educators often focus on preparing young people for work, but preparation for leisure is equally important. A person who enters adulthood without knowing how to use free time constructively is vulnerable to restlessness and dissatisfaction. Teaching the value of purposeful leisure is therefore an essential part of education for life.

Leisure also plays a quiet role in emotional health. Thoughtful use of free time provides balance. It offers relief from pressure without escape from responsibility. It gives space for reflection and renewal. People who use leisure well tend to approach work with greater energy and patience. They are less likely to feel trapped by routine because their lives contain variety and self-chosen meaning.

Over time, leisure choices accumulate. Hours become habits, and habits become character. A person may not notice change from week to week, but years reveal the difference. One path leads to growth, skill, and confidence. The other leads to restlessness and regret. The difference often lies not in talent or luck, but in how free time is treated.

The intelligent use of leisure time does not demand perfection or constant productivity. It demands awareness, balance, and honesty. It recognizes that rest and enjoyment are necessary, but that they are most satisfying when paired with growth. It accepts that not every activity must have a grand purpose, but that life as a whole benefits when leisure supports development rather than emptiness.

Ultimately, leisure time is a gift of modern life, but it is also a responsibility. It offers freedom unmatched by obligation, and with that freedom comes choice.

Whether that choice leads to boredom or fulfillment depends on how consciously it is made.

Time will pass regardless. The **Question** is not whether leisure exists, but whether it will be used wisely.

Those who learn to use leisure time well gain more than enjoyment. They gain skills, confidence, balance, and direction. They discover interests that shape their futures and habits that strengthen their character. They do not merely pass time; they invest it. And in doing so, they turn free hours into one of the most powerful tools for better living and lasting personal growth.

Questions & Answers

Question 1: Why does the chapter argue that leisure time reveals personal direction more clearly than work or obligation?

Answer: Work reflects what a person must do, often shaped by necessity, rules, or external demands. Leisure time, by contrast, is guided almost entirely by choice. Because it arrives without instruction, how a person uses it exposes values, interests, and long-term direction. The chapter suggests that leisure reveals identity in its purest form, showing who a person chooses to become when no one is watching or demanding performance.

Question 2: How does the chapter redefine boredom as a signal rather than a problem?
Answer: Boredom is presented not as a lack of activity, but as a lack of meaning. When leisure is unplanned or purposeless, time dissolves into restlessness. This discomfort is interpreted as a signal of unused potential and unclaimed freedom. Instead of blaming circumstances, the chapter encourages individuals to recognize boredom as an invitation to engage, choose, and develop direction through intentional action.

Question 3: In what way does self-directed leisure strengthen independence and emotional resilience?
Answer: When leisure depends entirely on others or external stimulation, frustration becomes inevitable as plans change or availability disappears. Self-directed leisure restores control by allowing enjoyment and growth regardless of circumstances. Through hobbies, learning, and personal projects, individuals develop patience, persistence, and confidence. This independence reduces vulnerability to boredom and emotional dissatisfaction, strengthening resilience over time.

Question 4: Why does the chapter emphasize contrast and change as essential functions of leisure?
Answer: Leisure provides renewal by offering contrast to daily obligations. Mental work is balanced by physical activity; repetitive tasks are refreshed by creativity. This contrast prevents fatigue from turning into discouragement and routine into stagnation. The chapter makes clear that leisure is not escape from life, but restoration within it—maintaining balance so effort remains sustainable and meaningful.

Question 5: How do repeated leisure choices gradually shape character and future outcomes?

Answer: The chapter highlights that leisure choices accumulate quietly over time. Hours become habits, habits become skills, and skills influence confidence, direction, and opportunity. While change may be invisible in the short term, years reveal the consequences of how free time is treated. The difference between growth and regret often lies not in talent or luck, but in whether leisure is used as distraction or as investment.

End of the Chapter

Chapter-9.
Social Acceptability and the Search for Belonging

Humans crave connection, yet true belonging comes from understanding and respect. This chapter examines how social awareness, empathy, and self-presentation affect acceptance. It guides readers to navigate societal expectations, form authentic bonds, and find their place without compromising individuality.

From the moment individuals begin to recognize themselves as separate beings, they also begin to sense their place among others. This awareness deepens with age, and by adolescence it becomes one of the most powerful influences on thought, emotion, and behavior.

The desire to belong is not a weakness, nor is it a shallow craving for approval. It is a fundamental human need rooted in survival, identity, and emotional security.

To be accepted is to feel seen, to feel legitimate, and to feel anchored in the human world. Social acceptability, therefore, reaches far beyond popularity. It shapes self-image, confidence, and the way individuals interpret their value within society.

During adolescence, social awareness intensifies. Physical growth, emotional sensitivity, and expanding independence converge at the same time, creating a period of heightened self-consciousness. School becomes more than an academic environment; it becomes a social ecosystem governed by unwritten rules and subtle hierarchies. Friendships, rivalries, alliances, and reputations form rapidly. Within this setting, acceptance can feel like confirmation of worth, while exclusion can feel like a personal judgment. These experiences are rarely neutral. They leave impressions that influence personality development long after adolescence has passed.

Social groups emerge naturally in any setting where people spend time together. These groups are often formed around shared interests, confidence, appearance, social skills, or perceived status. Some individuals move easily into the center of these groups, navigating social interaction with apparent ease. Others remain at the margins, watching, hoping, and hesitating. Still others deliberately choose distance, finding satisfaction in independence rather than group belonging. Each position carries emotional meaning, and none is accidental. Social placement is the result of both personal traits and environmental influences.

Those who are widely accepted often seem confident, relaxed, and socially skilled. Their comfort reinforces positive feedback from others, creating a cycle in which acceptance breeds confidence and confidence attracts acceptance. Yet this ease is not always innate. Often, it is learned through experience, observation, and practice. Socially accepted individuals tend to understand the unspoken language of interaction—when to speak, when to listen, how to show interest, and how to adapt to different personalities. These skills, while rarely acknowledged, are powerful tools for connection.

At the opposite end of the social spectrum are those who struggle to find acceptance. They may possess intelligence, kindness, creativity, or integrity, yet still feel invisible or excluded. Their difficulty often lies not in their qualities, but in their self-perception. Fear of rejection, embarrassment, or judgment can inhibit natural expression. This hesitation may appear as shyness, withdrawal, or guarded behavior, which others may misinterpret as disinterest or aloofness. Thus, a cycle begins in which fear limits engagement, and limited engagement reinforces exclusion.

Between these extremes lies a broad middle ground occupied by individuals who are socially involved but emotionally uncertain. They have friends, participate in group life, and enjoy moments of belonging, yet remain sensitive to approval. Their confidence fluctuates. Praise reassures them, while criticism or exclusion unsettles them deeply. For these individuals, social acceptability feels conditional—something that must be maintained rather than assumed. A single overlooked invitation or dismissive comment can trigger doubt, even when no harm was intended.

There are also those who consciously choose independence. These individuals do not reject others, but they do not measure themselves by group approval. Their interests may differ from those of their peers, or their temperament may favor solitude and focused pursuits. They find fulfillment in personal growth, creativity, or intellectual exploration. While this path can be healthy and authentic, it is often misunderstood. Others may mistake independence for rejection or indifference. Yet such individuals demonstrate that belonging can exist without constant validation.

Social acceptability is influenced by many factors, some within an individual's control and others shaped by circumstance. Family environment plays a significant role. Children absorb social attitudes from parents long before they understand them consciously. Parents who are comfortable in social settings often model confidence, conversation, and hospitality. Those who are anxious or withdrawn may unintentionally model caution and self-doubt. These patterns are not deliberate, but they influence how young people approach social life.

Economic circumstances and cultural expectations also shape social experience. Adolescents are acutely aware of differences in clothing, housing, activities, and opportunities. These differences can affect confidence and inclusion, even when they are not openly discussed. Young people may assume that acceptance depends on external markers of status, leading them to feel inadequate or excluded when they do not match perceived norms. Such assumptions can quietly undermine self-esteem.

Adolescence is a particularly vulnerable period because identity is still forming. Young people look to peers for comparison and validation. Inclusion becomes symbolic, representing belonging and worth.

An invitation can feel like confirmation of value, while rejection can feel like a verdict. In reality, these moments are rarely as decisive as they seem. Social dynamics are fluid, influenced by chance, timing, and circumstance. Yet at the time, their emotional impact is profound.

Rejection, especially when repeated, can leave lasting impressions. It does not merely disappoint; it teaches. It may teach a young person to expect exclusion, to lower expectations, or to doubt their own worth. Over time, these lessons can shape behavior and personality. A person who anticipates rejection may stop reaching out, avoid social situations, or adopt defensive attitudes. These patterns, once established, can persist into adulthood unless consciously addressed.

Yet rejection does not determine destiny. Social acceptability is not fixed.

It changes with environment, maturity, and self-awareness. Many individuals who feel overlooked or excluded in adolescence later find belonging in new contexts where their qualities are recognized and valued.

Adult life offers greater diversity of environments and relationships, allowing individuals to find communities aligned with their interests and values. What once seemed essential often loses its power as perspective grows.

One of the most important truths about social acceptance is that it rarely depends on a single trait. It is not guaranteed by appearance, wealth, intelligence, or talent alone. Rather, it emerges from a complex interaction of confidence, communication, empathy, and adaptability. Individuals who learn to read social situations, respond thoughtfully, and express themselves comfortably tend to find greater acceptance, regardless of background. These skills are not innate gifts; they can be learned and practiced.

This is why social skills matter. Social acceptability is not about pretending to be someone else. It is about learning how to present oneself authentically while remaining sensitive to others. Skills such as listening attentively, showing genuine interest, managing anxiety, and expressing warmth create connection. They are not manipulative techniques; they are expressions of respect. When young people understand this, social life becomes less mysterious and less threatening.

Parents play a crucial role in supporting this understanding. Their attitudes toward social life influence how children interpret their experiences. Parents who dismiss social concerns as trivial may unintentionally invalidate genuine emotional pain. Conversely, parents who treat social acceptance as the ultimate measure of success may increase pressure and anxiety. Balance is essential. Young people need reassurance that social experiences matter, but also perspective that they do not define worth.

Supportive parents help by creating environments where social interaction feels safe and manageable. This does not require elaborate events or material display.

Often, it simply means openness—allowing young people to bring friends home, encouraging conversation, and modeling hospitality without perfectionism.

Informal, relaxed settings teach that connection does not require performance, only sincerity.

Parents also help by sharing their own experiences honestly. Many adults forget how vulnerable adolescence felt. Remembering and acknowledging past insecurities builds empathy. When young people see that adults once struggled too—and survived—they gain hope.

They learn that awkwardness is not permanent and that confidence grows through experience, not avoidance.

Equally important is teaching resilience. Social life inevitably involves disappointment. Not every effort leads to acceptance. Teaching resilience does not mean minimizing pain; it means helping young people interpret experiences realistically. A missed invitation is not a final judgment. A rejection is not proof of unworthiness. Learning to recover, reflect, and try again is one of the most valuable social skills of all.

It is also vital to broaden the definition of success. Popularity is visible, but it is not the only form of acceptance. Deep friendships, mutual respect, and shared values often provide more lasting satisfaction than widespread recognition. Encouraging young people to invest in genuine connections rather than chasing approval protects emotional well-being and fosters healthier relationships.

Social acceptability evolves with age. The traits that dominate adolescent social hierarchies often lose importance in adulthood. Confidence remains valuable, but it becomes paired with reliability, empathy, and competence.

Many individuals who felt overlooked in youth find that their strengths—persistence, thoughtfulness, integrity—are highly valued later in life. Time often reshapes social landscapes in unexpected ways.

Educational environments and communities also influence social acceptance. Settings that reward cooperation rather than competition tend to reduce exclusion.

When diverse talents are recognized, more individuals find a sense of belonging. Inclusive activities, mixed groups, and shared responsibilities soften rigid boundaries and teach that belonging is not limited to a single circle.

Ultimately, social acceptability matters because humans are social beings. We are shaped by interaction. Acceptance reassures us that we are seen and valued. Yet it must be balanced with self-acceptance. When individuals learn to value themselves independently of external approval, social experiences become less threatening and more enriching. They approach others not from desperation, but from openness.

The goal is not universal popularity, but social confidence—the ability to engage without fear, to handle rejection without collapse, and to enjoy connection without losing oneself. This confidence allows individuals to move through social environments with flexibility and dignity. Acceptance becomes welcome, but not essential for self-worth.

Social acceptability is both important and incomplete. It influences development, shapes personality, and leaves emotional traces. But it is not destiny. With time, support, and self-awareness, individuals can grow beyond early limitations. They can learn to create connection rather than wait for it, to value depth over display, and to carry a sense of worth that no invitation can grant or remove.

In the end, belonging begins within. When individuals understand their own value, social life becomes a field of opportunity rather than judgment. Acceptance becomes something to share, not something to beg for. And it is this quiet shift—toward self-respect, empathy, and resilience—that allows social experience to enrich life rather than define it.

Questions & Answers

Question 1: Why does the chapter describe the desire for social acceptability as a fundamental human need rather than a weakness?

Answer: The chapter makes clear that the desire to belong is rooted in survival, identity, and emotional security. Acceptance reassures individuals that they are seen, legitimate, and anchored in the human world. Rather than being a shallow craving for approval, social acceptability shapes self-image and confidence from an early age. By framing it as a basic human need, the chapter removes moral judgment and instead encourages understanding of how deeply social experiences influence development.

Question 2: How do adolescent social experiences leave lasting impressions even when they are temporary or situational?

Answer: Adolescence is a period of heightened sensitivity because identity is still forming. Social experiences during this time are often interpreted as judgments of worth rather than as situational events. Acceptance can feel like validation, while exclusion can feel definitive and personal. Although social dynamics are fluid, repeated experiences of rejection or inclusion can teach expectations that shape behavior, confidence, and relationships long after adolescence has passed.

Question 3: Why does fear of rejection often create a self-reinforcing cycle of social exclusion?

Answer: The chapter explains that fear of rejection inhibits natural expression. This hesitation may appear as withdrawal, guardedness, or silence, which others may misread as disinterest or aloofness. Reduced engagement then limits opportunities for connection, reinforcing exclusion and confirming the original fear. Over time, this cycle can solidify into a pattern unless consciously addressed through awareness and gradual re-engagement.

Question 4: In what way does the chapter challenge the idea that social acceptability depends on fixed traits like appearance or status?
Answer: Rather than attributing acceptance to a single trait, the chapter emphasizes a combination of learnable skills: communication, empathy, adaptability, and confidence. Social acceptability is shown to be dynamic and context-dependent, changing with environment and maturity. This perspective shifts social success from fate to growth, suggesting that individuals can improve their social experiences through practice and self-awareness rather than self-rejection.

Question 5: How does the chapter reconcile the importance of belonging with the necessity of self-acceptance?
Answer: The chapter concludes that while acceptance matters, it must not become the sole source of self-worth. When individuals value themselves independently of approval, social interactions become less threatening and more genuine. Belonging then becomes something to share rather than to seek desperately. This balance allows individuals to engage socially with dignity, resilience, and openness, ensuring that social life enriches identity rather than defines it.

End of the Chapter

Chapter–10.
Manners in Society: The Foundation of a Healthy, Happy, and Successful Life

Manners are the silent framework of social harmony. This chapter highlights how etiquette, consideration, and respect foster trust, reduce conflict, and create positive impressions. Practicing good manners isn't superficial—it's a cornerstone of happiness, strong relationships, and lasting success in every aspect of life.

From the earliest moments of social awareness, human beings begin learning how to live among others. Even before language becomes refined or logic fully develops, children sense approval, disapproval, warmth, and rejection through everyday interactions. These early impressions quietly shape patterns of behavior that often last a lifetime. Long before academic achievement, professional competence, or social status enter the picture, the way a person speaks, listens, reacts, and responds begins to define how they fit into the world. These patterns, refined over time, are what we call manners. They are not superficial gestures or rigid formalities; they are the living expression of respect, empathy, and self-awareness that allow human beings to coexist with dignity.

Manners are sometimes misunderstood as a set of artificial rules meant to restrict natural behavior. In reality, they exist to protect something far more valuable than comfort or tradition: relationships. Wherever people must share space, time, or responsibility, manners act as a stabilizing force. They prevent unnecessary conflict, soften differences, and create an atmosphere in which cooperation becomes possible. A society governed only by laws may enforce order, but without manners it cannot sustain goodwill. Laws restrain behavior; manners guide it. One works through fear of consequence, the other through understanding.

Healthy social living begins with awareness. Awareness of others as individuals with thoughts, feelings, pressures, and dignity equal to one's own. Every public space, every workplace, every neighborhood is a shared environment. Manners arise from recognizing that one's actions ripple outward. A careless word, a dismissive tone, or a selfish act may seem insignificant in isolation, yet repeated daily across a society, such behaviors create tension, mistrust, and emotional distance. Conversely, small acts of courtesy accumulate into trust, ease, and mutual respect.

In everyday life, manners guide countless interactions that often go unnoticed precisely because they work so well. Greeting others politely acknowledges their presence. Waiting one's turn affirms fairness. Listening attentively communicates value. Apologizing restores balance when mistakes occur. These actions are simple, yet their impact is profound. They reduce friction and allow people to move through life with less anxiety and defensiveness. When manners are present, social life feels smoother and safer. When they are absent, even routine interactions can feel exhausting.

Respect for authority and responsibility is one of the cornerstones of social manners. Every functioning society depends on individuals who carry specific roles for the collective good. Teachers educate, caregivers nurture, workers provide essential services, and leaders coordinate shared efforts. Showing respect toward these roles is not submission or weakness; it is recognition of interdependence. Disrespect erodes cooperation and undermines trust. Respectful conduct, by contrast, strengthens both institutions and interpersonal relationships, reminding individuals that no one operates in isolation.

Listening deserves special attention as a social virtue. Listening is not merely remaining silent while another speaks; it is an active process of attention and understanding. Many conflicts persist not because of irreconcilable differences, but because people feel unheard. When individuals listen sincerely, tensions often ease before solutions are even discussed. A society in which listening is valued becomes more resilient, because it allows disagreement without dehumanization. Differences can exist without turning into divisions.

Consideration for others' feelings lies at the heart of manners. Words spoken casually can wound deeply if they disregard sensitivity or context. Humor, criticism, or teasing that ignores another person's vulnerability may be defended as harmless, yet its emotional impact can linger long after the moment has passed. Manners encourage restraint not to suppress expression, but to ensure that expression does not cause unnecessary harm. Empathy—the ability to imagine oneself in another's position—is the emotional engine of good manners.

Accountability is another essential dimension of social behavior. In any shared life, mistakes are inevitable. What matters is how individuals respond to them.

A sincere acknowledgment of error and a genuine apology can restore trust and dignity.

Apologies are not admissions of weakness; they are demonstrations of maturity and responsibility. In a society where people refuse to admit mistakes, resentment accumulates and relationships deteriorate. In a society where accountability is practiced, healing becomes possible.

Fairness, closely tied to manners, sustains social confidence. Taking turns, following shared rules, and doing one's share ensure that no individual benefits at the expense of others. When fairness is perceived, people are more willing to cooperate and contribute. When it is absent, frustration and suspicion arise. Manners protect fairness by encouraging self-restraint and mutual respect, reminding individuals that personal convenience must sometimes yield to collective good.

Order and responsibility also shape the social environment. Punctuality, cleanliness in shared spaces, and reliability in commitments demonstrate respect for others' time and effort. These habits may appear minor, yet they have cumulative effects.

A person who consistently neglects responsibilities places an unfair burden on others. Manners remind individuals that freedom carries obligation, and that consideration is the price of shared life.

Self-control is another quiet gift of manners. Society presents constant opportunities for irritation, impatience, and conflict. Manners provide a framework for responding thoughtfully rather than impulsively. Self-control does not deny emotion; it channels it constructively. A society that values restraint alongside expression maintains balance. Without it, freedom becomes chaos and assertiveness becomes aggression.

The development of manners does not occur automatically. It requires observation, guidance, and practice. People learn how to behave largely by watching those around them, especially those they admire or depend upon. Families, schools, and communities all play formative roles. When courtesy, patience, and respect are consistently modeled, they become internalized. When rudeness, mockery, or indifference are tolerated or rewarded, they spread just as quickly.

Parents and educators carry particular responsibility in this process. Instruction alone is insufficient; example is decisive. Young people learn far more from what adults do than from what they say. When adults demonstrate kindness under pressure, respect during disagreement, and accountability after mistakes, they teach manners without a single lecture. Conversely, when adults display contempt, impatience, or disregard, those behaviors are absorbed just as readily.

Manners are inseparable from personal success. Achievement in any field depends on relationships. Cooperation, trust, and communication are essential to sustained success. A person may possess exceptional skill or intelligence, but without manners those qualities often fail to translate into lasting opportunity. Manners open doors that talent alone cannot. More importantly, they keep those doors open. Over time, character proves more reliable than ability.

Social confidence grows naturally from good manners. Knowing how to behave in various situations reduces anxiety and uncertainty. A person who understands social expectations moves through unfamiliar environments with greater ease. This confidence is not rooted in superiority, but in preparedness. Manners provide guidance when circumstances are unclear, allowing individuals to respond with grace rather than fear.

Manners also support emotional health. Respectful interactions reduce stress and create psychological safety. In environments where people feel valued, they are more likely to express themselves honestly and listen openly.

This openness strengthens relationships and communities alike. Emotional safety is not a luxury; it is a foundation for well-being. A happy society is not one without disagreement, but one in which disagreement is managed with respect. Manners allow people to challenge ideas without attacking individuals. They keep discussion focused on substance rather than ego. In doing so, they protect both freedom of expression and human dignity. Without manners, debate becomes hostility. With manners, it becomes progress.

While manners evolve with culture and time, their core purpose remains unchanged. Politeness, empathy, fairness, responsibility, and respect adapt to new contexts, technologies, and social structures, yet their essence endures.

Manners are living principles, not static rules. They change in form, but not in function.

Learning manners is a lifelong process. No one practices them perfectly. What matters is intention and effort. Each interaction presents an opportunity to improve. Each misstep offers a chance to reflect and grow.

A society that values growth over perfection encourages better behavior and stronger bonds.

Ultimately, manners are expressions of character. They reveal how a person views others and themselves. A person with good manners demonstrates self-respect as well as respect for the world around them. They understand that personal freedom and social responsibility are not opposites, but partners. One cannot exist meaningfully without the other.

A healthy society depends on manners to function smoothly. A happy society depends on them to maintain trust and goodwill. A successful life depends on them to build relationships, earn respect, and navigate challenges with dignity. Manners are not decorative additions to life; they are its foundation.

In society, as in all shared spaces, manners guide behavior when no one is watching and character is quietly revealed. They shape daily experiences and long-term outcomes alike. By practicing manners consistently—at home, at work, in public, and in private—individuals contribute not only to their own growth, but to the health and happiness of the world they share.

In the end, manners are not about following rules for their own sake. They are about choosing consideration over convenience, respect over impulse, and understanding over indifference. They are the daily choices that turn coexistence into community, interaction into connection, and ordinary life into a healthy, happy, and successful human journey.

Questions & Answers

Question 1: Why does the chapter describe manners as a foundation of society rather than as social decoration or etiquette?

Answer: The chapter emphasizes that manners are not superficial gestures meant to impress, but fundamental behaviors that protect relationships and social stability. Laws may enforce order through fear of punishment, but manners sustain goodwill through understanding and empathy. Without manners, society may function mechanically, but it cannot flourish emotionally. Manners form the invisible structure that allows cooperation, trust, and dignity to exist in everyday interactions.

Question 2: How does the chapter distinguish between obedience to rules and genuine social responsibility?

Answer: Obedience relies on external control, while social responsibility arises from internal awareness. The chapter explains that manners are rooted in recognizing the impact of one's actions on others. A person who follows rules out of fear may comply only when watched, but a person guided by manners acts thoughtfully even when unobserved. This distinction shows that manners reflect character, not compliance, and reveal how individuals behave when authority is absent.

Question 3: Why are listening and empathy presented as central social virtues rather than optional personal traits?
Answer: Listening and empathy are described as essential because many social conflicts persist not due to disagreement, but due to feeling unseen or unheard. Active listening affirms dignity and reduces emotional tension before solutions are even discussed. Empathy allows individuals to regulate expression so that honesty does not become harm. Together, these qualities prevent dehumanization and allow disagreement to exist without destroying relationships, making them pillars of a resilient society.

Question 4: In what way does accountability transform mistakes from social threats into opportunities for trust?
Answer: The chapter explains that mistakes are inevitable in shared life; what matters is response. Accountability—acknowledging error and offering sincere apology—restores balance and dignity. Rather than weakening authority or respect, accountability strengthens trust by demonstrating maturity and responsibility. Societies that reject accountability accumulate resentment, while those that practice it maintain the possibility of repair and growth.

Question 5: How does the chapter connect manners to long-term personal success rather than short-term social approval?
Answer: The chapter argues that sustained success depends on relationships, not talent alone. Manners enable cooperation, trust, and communication, which keep opportunities open over time. While skill may attract attention, character determines whether respect endures. Manners also provide social confidence, reduce anxiety, and support emotional health, making them essential not just for achievement, but for a balanced and fulfilling life.

End of the Chapter

Chapter–11.
Improve Your Reading Habits: The Path to an Influential, Educated, and Successful Life

Reading is a powerful tool for growth and influence. This chapter shows how cultivating effective reading habits expands knowledge, sharpens thinking, and strengthens communication. By approaching reading with purpose and focus, you can unlock personal development and gain the skills needed to thrive in life and career.

From the earliest years of formal learning, reading stands at the center of human development, yet its true power is often underestimated.

Many people associate reading only with examinations, textbooks, or compulsory assignments, failing to recognize that it is one of the most decisive forces shaping the quality of thought, communication, and life itself.

Reading is not merely a method of gathering information; it is a discipline that trains the mind, expands awareness, refines language, and strengthens judgment. Long after classrooms are left behind, reading continues to influence how individuals interpret the world, respond to challenges, and shape their personal and professional identities.

An educated life does not begin with certificates or degrees. It begins with the habit of engaging deeply with ideas. Reading provides access to the accumulated knowledge, experience, and wisdom of countless minds across time. Each book, article, or thoughtful text represents a conversation extended beyond its original moment. When a person reads attentively, they enter that conversation, learning not only facts, but ways of thinking. Over time, this repeated engagement develops intellectual maturity and clarity that no shortcut can replace.

Many individuals believe they read well simply because they can read quickly or finish large volumes of material. Speed alone, however, is not understanding. Reading without focus is little more than passing time.

True reading requires presence of mind. It demands attention, curiosity, and willingness to slow down when ideas become complex. When reading becomes deliberate rather than rushed, it begins to shape reasoning and insight. The reader no longer chases pages; they absorb meaning.

Purpose transforms reading from habit into power. When a reader approaches a text with intent, they read differently. They look for structure, argument, and implication. They ask what the author is trying to convey and why.

This mental engagement strengthens comprehension and retention. Instead of memorizing isolated facts, the reader grasps relationships between ideas. This ability to see connections is one of the hallmarks of an educated mind.

Vocabulary growth is one of the most visible benefits of consistent reading, yet it is also one of the most misunderstood. Vocabulary is not about displaying intelligence or sounding impressive. It is about precision of thought. Words shape how ideas are formed and expressed.

A limited vocabulary restricts expression, often leading to frustration and misunderstanding. Reading introduces words naturally, in context, allowing the reader to absorb not only definitions but emotional tone, nuance, and proper usage. Over time, language becomes more flexible, expressive, and confident.

This improvement in language directly enhances communication. People who read regularly tend to speak more clearly, write more persuasively, and listen more attentively. They understand how ideas are organized and how arguments unfold. As a result, they are better equipped to explain themselves, negotiate differences, and influence others constructively. In a world where communication often determines opportunity, this skill is invaluable.

Confidence is another quiet gift of strong reading habits. Confidence does not arise from knowing everything, but from knowing how to learn. A person who reads effectively trusts their ability to understand new material. They are less intimidated by unfamiliar topics because they possess the tools to explore them. This confidence extends into academic, professional, and social settings, allowing individuals to participate actively rather than withdraw in uncertainty.

Reading also trains attention, a skill increasingly rare in a world filled with distraction. Sustained reading requires concentration over time. By choosing to read regularly, individuals strengthen their ability to focus, resist impulse, and remain mentally engaged.

This discipline benefits every area of life, from problem-solving to emotional regulation. A focused mind is calmer, more deliberate, and more capable of thoughtful response.

Patience is another virtue cultivated through reading. Complex ideas do not always yield immediate understanding. Some passages must be reread, questioned, and reflected upon. This process teaches persistence and humility. Readers learn that confusion is not failure, but part of learning. They become comfortable with uncertainty, trusting that clarity emerges through effort. This mindset prepares individuals to face challenges beyond books, where solutions are rarely instant.

Reflection deepens the value of reading. Information that is not reflected upon fades quickly. Thoughtful readers pause to consider what they have read, how it relates to prior knowledge, and what questions it raises. This reflection transforms reading into personal insight. It encourages readers to examine assumptions, reconsider opinions, and refine beliefs. Over time, this habit of reflection leads to wiser decision-making and stronger judgment.

Note-taking, when used thoughtfully, strengthens this process further. Writing down key ideas, unfamiliar terms, or personal reactions creates an active relationship with the text. Notes help organize information and make later review meaningful. They turn reading into dialogue rather than passive consumption. Over time, this practice sharpens analytical ability and reinforces memory.

Reading also shapes values and emotional intelligence. Through exposure to diverse perspectives, readers encounter lives, cultures, and struggles different from their own. This exposure broadens understanding and fosters empathy. Readers learn that human experience is complex and multifaceted. They become more tolerant, less rigid, and more capable of seeing beyond their immediate circumstances. These qualities are essential for leadership, cooperation, and meaningful relationships.

Ethical awareness is often strengthened through reading. Stories and essays present moral dilemmas, conflicting values, and difficult choices. Readers are invited to consider consequences, motivations, and responsibilities. This engagement develops moral reasoning, encouraging individuals to think beyond convenience or impulse. An educated life is not defined solely by knowledge, but by wisdom—the ability to apply understanding responsibly.

In professional life, reading habits provide a decisive advantage. Knowledge evolves continuously, and skills quickly become outdated. Those who read regularly remain adaptable. They stay informed, recognize trends, and understand context.

Reading enables individuals to learn independently, reducing reliance on formal instruction. Lifelong readers remain relevant because they remain learners.

Reading also supports emotional well-being. It offers both stimulation and rest. A meaningful text engages the mind while providing relief from constant external pressure. Reading creates space for thought, allowing individuals to step back from immediate concerns and gain perspective. This balance supports mental health, resilience, and emotional stability.

Consistency is more important than intensity in developing reading habits. Reading a small amount daily builds momentum more effectively than occasional bursts of effort.

Over time, reading becomes integrated into daily life, no longer experienced as a task but as a natural activity. This consistency transforms reading from something one does into part of who one is.

Choosing reading material thoughtfully also matters. While entertainment has its place, reading for growth requires intention. Material that challenges thinking, improves language, or deepens understanding offers lasting value.

A balanced reading life includes both enjoyment and enrichment, allowing the reader to grow without losing pleasure.

Improving reading habits does not require perfection. Everyone experiences distraction and fatigue. What matters is awareness and persistence.

Returning attention to the text, rereading when necessary, and reflecting afterward all strengthen the habit. Each small improvement builds skill and confidence.

Education does not end with graduation. Reading ensures that learning continues throughout life. It keeps the mind active, curious, and engaged with the world. Those who read remain open to growth, capable of change, and prepared for complexity.

Ultimately, reading habits shape not only what a person knows, but who they become.

They influence communication, confidence, judgment, and values. A thoughtful reader develops depth, clarity, and perspective.

They gain not only information, but insight—the ability to see meaning beneath surface detail.

An influential, educated, and successful life is not built suddenly. It is built quietly, page by page, idea by idea. Reading is the habit that sustains this growth. By improving reading habits, individuals invest in clearer thinking, stronger communication, and a richer human experience. In doing so, they equip themselves not merely to succeed, but to live thoughtfully, responsibly, and with enduring purpose.

Questions & Answers

Question 1: Why does the chapter argue that reading is a discipline of thought rather than merely a method of acquiring information?

Answer: The chapter emphasizes that reading trains how the mind works, not just what it knows. True reading requires attention, patience, reflection, and engagement with ideas. Through this process, readers learn to analyze arguments, recognize connections, and refine judgment. Information alone can be forgotten, but disciplined reading reshapes thinking patterns, influencing how individuals interpret the world and make decisions long after the material itself is set aside.

Question 2: How does purposeful reading transform understanding more effectively than speed or volume?

Answer: Reading quickly or consuming large amounts of material does not guarantee comprehension. Purposeful reading involves approaching a text with intent—seeking meaning, structure, and implication. This mindset encourages readers to slow down when necessary, ask questions, and reflect on ideas. As a result, understanding becomes deeper and more durable, allowing readers to grasp relationships between concepts rather than memorizing isolated facts.

Question 3: Why is vocabulary described as a foundation for clear thinking rather than a tool for display?

Answer: The chapter explains that vocabulary shapes thought itself. Without precise words, ideas remain vague and difficult to express. Reading introduces language in context, allowing readers to absorb nuance, tone, and meaning naturally. As vocabulary grows, so does the ability to think clearly, communicate effectively, and avoid frustration or misunderstanding. Thus, vocabulary is presented as a means of intellectual precision, not social superiority.

Question 4: In what way do reading habits cultivate confidence and emotional stability?
Answer: Confidence arises not from knowing everything, but from trusting one's ability to learn. Regular readers become comfortable with unfamiliar topics because they possess the skills to understand them. Reading also strengthens attention and patience, helping individuals manage distraction and emotional impulse. By fostering focus, reflection, and perspective, reading contributes to emotional steadiness and reduces anxiety when facing complexity or uncertainty.

Question 5: How does the chapter connect consistent reading habits to long-term influence and success?
Answer: The chapter argues that influence and success depend on communication, adaptability, judgment, and values—all of which are strengthened through reading. Consistent reading keeps individuals intellectually flexible, informed, and capable of independent learning. Over time, small daily reading habits accumulate into clarity of thought, persuasive communication, and ethical awareness. This gradual development, rather than sudden achievement, forms the foundation of a meaningful, educated, and influential life.

End of the Chapter

Chapter–12.
Keeping Up With Your Studies: Discipline, Focus, and the Power of Intelligent Reading

Success in learning comes from more than effort—it comes from strategy. This chapter emphasizes discipline, concentration, and smart reading techniques. By mastering these practices, students and lifelong learners can retain knowledge, excel academically, and develop the mental agility necessary to achieve both personal and professional goals.

There comes a quiet turning point in every learner's life when effort begins to matter more than intention. Until that moment, it is easy to believe that attending classes, owning textbooks, and wishing for good results are enough.

But gradually, reality reveals something deeper: progress belongs to those who manage their studies with awareness, discipline, and purpose. Keeping up with studies is not about pressure or punishment; it is about learning how to direct one's energy intelligently so that learning becomes steady, meaningful, and sustainable.

Many students fall behind not because they lack intelligence, but because they underestimate the value of daily attention. Falling behind rarely happens all at once. It develops through small acts of delay, scattered focus, and misplaced priorities. A missed assignment here, a distracted study period there, and soon confusion replaces clarity. What was once manageable begins to feel heavy. At that point, studying stops being a routine and becomes a source of stress. Understanding this process is the first step toward reversing it.

At the heart of steady academic progress lies a simple truth: time is neutral, but how it is used determines outcomes. Every student is given similar hours in a day and similar opportunities within school. What separates those who keep up from those who struggle is not extra time, but effective use of time. Productive study does not require endless hours; it requires focused, intentional periods of work. Short, well-directed study sessions often achieve more than long hours filled with distraction.

Distraction is one of the greatest obstacles to consistent learning. Noise, interruptions, and divided attention weaken concentration and fragment understanding. Studying in an environment that invites distraction turns learning into a battle. A quiet, orderly space, by contrast, supports mental focus and reduces fatigue. When the mind is not constantly pulled in different directions, it can settle into deeper understanding. Preparing the study environment is not a minor detail; it is a strategic decision.

Mental readiness is just as important as physical environment. Sitting down to study without knowing what needs to be done wastes valuable time. Keeping up with studies requires clarity. Assignments must be written down, understood, and prioritized. When tasks are vague, the mind hesitates. When tasks are clear, action follows naturally. Clarity transforms responsibility into manageable steps, making progress possible.

Effective students learn to use study time at school wisely. Study periods are not breaks; they are opportunities. When used well, they reduce the workload later and prevent assignments from piling up. Completing even part of an assignment during school hours creates momentum. Ignoring this time often results in longer evenings filled with pressure and fatigue. School study time, when respected, becomes one of the strongest allies of steady progress.

Organization quietly shapes success. Disorganized notebooks, scattered papers, and missing supplies create unnecessary resistance. Each moment spent searching for information is a moment stolen from learning. Organizing materials by subject, keeping assignments recorded, and maintaining basic supplies creates a system that supports efficiency. Order in materials encourages order in thought, and order in thought supports understanding.

Understanding, not mere completion, is the real goal of study. Reading assignments without grasping meaning leads to repeated struggle. When unfamiliar words or concepts appear, they must be addressed immediately rather than ignored. Looking up definitions, reviewing notes, and asking questions prevent confusion from spreading. Time spent understanding saves time later by reducing repetition and frustration.

Reading plays a central role in keeping up with studies. Reading is not simply about finishing pages; it is about engaging with ideas. Effective reading requires attention to main ideas, structure, and meaning. Skilled readers do not rush blindly through material. They read with purpose, noticing key points and connections. This kind of reading transforms information into knowledge and knowledge into insight.

Strong reading habits also build vocabulary, which directly influences comprehension and communication. Words are tools of thought. The richer one's vocabulary, the more precisely ideas can be understood and expressed. Reading exposes learners to language in context, teaching nuance and clarity naturally. Over time, this improves both academic performance and everyday communication.

Reflection deepens the value of reading and study. Information that is not reflected upon fades quickly. Pausing to consider what has been read, how it connects to prior knowledge, and what questions remain transforms study into learning. Reflection strengthens memory and encourages critical thinking. It turns passive reading into active understanding.

Note-taking supports this process by creating a bridge between reading and thinking. Writing down key ideas, questions, or summaries reinforces comprehension. Notes are not meant to copy text, but to capture meaning. Over time, thoughtful note-taking sharpens attention and improves recall, making future study more efficient.

Planning study time is another essential habit. Not all subjects demand equal effort. Challenging tasks should be addressed first, when concentration is strongest. Easier work can follow. This approach reduces procrastination and builds confidence early in a study session. Planning also ensures that no subject is neglected, preventing last-minute panic.

Keeping up with studies does not mean working endlessly without rest. Breaks, when used wisely, support concentration and prevent burnout. Short pauses allow the mind to recover and return refreshed. The key is balance. Breaks should restore focus, not replace effort. Learning to work, pause, and resume efficiently is part of disciplined study.

Consistency is the true foundation of academic success. Studying a little every day is far more effective than attempting to recover lost ground in a single effort. Daily review reinforces memory, deepens understanding, and reduces anxiety. When study becomes a regular habit rather than an emergency response, it loses much of its emotional weight. Progress feels steady and achievable.

Responsibility is central to keeping up with studies. Teachers provide guidance, but learners must take ownership of their work. Waiting for reminders or excuses undermines growth. Accepting responsibility builds independence, one of the most valuable outcomes of education. This independence extends beyond school, preparing individuals for adult responsibilities.

Keeping up with studies also protects personal freedom. When work is handled responsibly, time opens up for recreation, friendships, and rest. Falling behind restricts freedom, replacing enjoyment with pressure. Discipline creates choice; neglect removes it. Understanding this connection changes the way students view study—from punishment to protection.

Improvement does not require perfection. Everyone falls behind at times. What matters is response. Recognizing difficulty early and acting immediately prevents small setbacks from becoming major obstacles. Even modest effort restores momentum. Progress builds motivation, and motivation fuels further progress.

Support systems matter as well. Parents, teachers, and peers can offer guidance and encouragement. Seeking help is not weakness; it is wisdom. However, support is most effective when combined with personal effort. Others can assist, but responsibility cannot be transferred.

The habits developed through consistent study and thoughtful reading shape life beyond academics. Time management, focus, perseverance, and clarity of thought are life skills. Individuals who learn to manage their studies effectively are better prepared to manage work, relationships, and responsibilities later on. Education becomes training not only in knowledge, but in character.

Confidence grows naturally when studies are handled well. Understanding replaces confusion. Preparation replaces anxiety. Effort replaces avoidance. This confidence is quiet and stable, grounded in experience rather than hope. It allows learners to face challenges with resilience instead of fear.

In the end, keeping up with studies is not about racing ahead or competing with others. It is about steady progress, honoring commitments, and respecting one's own potential. It is the daily practice of showing up prepared, attentive, and willing to learn. Through this practice, success reveals itself not as sudden brilliance, but as the result of disciplined, thoughtful effort.

When studies are kept in balance, life itself feels more balanced. Learning becomes meaningful rather than stressful. Work and enjoyment coexist without conflict. Keeping up with studies, supported by strong reading habits, is not merely an academic goal. It is a way of learning how to live with purpose, clarity, and confidence—qualities that serve a person well long after the classroom is left behind.

Questions & Answers

Question 1: Why does the chapter emphasize daily attention over intelligence as the key factor in academic progress?
Answer: The chapter makes clear that most students fall behind not because they lack ability, but because they underestimate the cumulative power of small delays and distractions. Progress erodes gradually through missed assignments, unfocused study sessions, and unclear priorities. Daily attention prevents this slow decline by maintaining continuity. Intelligence may determine potential, but consistent attention determines whether that potential is realized.

Question 2: How does clarity of tasks transform studying from stress into manageable effort?
Answer: Studying becomes overwhelming when responsibilities are vague. When tasks are clearly identified, written down, and prioritized, the mind moves naturally toward action instead of avoidance. The chapter highlights that clarity reduces hesitation and mental resistance. By breaking responsibilities into defined steps, studying shifts from an emotional burden into a practical process, restoring a sense of control.

Question 3: Why is intelligent reading presented as central to keeping up with studies rather than as a separate skill?
Answer: The chapter positions reading as the gateway to understanding in every subject. Simply completing pages does not prevent confusion; engaging with ideas does. Intelligent

reading—focused on structure, meaning, and key points—turns information into insight. Strong reading habits also build vocabulary, which directly improves comprehension and communication. Without intelligent reading, effort multiplies while results diminish.

Question 4: In what way does discipline protect personal freedom rather than restrict it?
Answer: Discipline is often mistaken for limitation, but the chapter reframes it as protection of freedom. When studies are handled responsibly, time opens for rest, relationships, and leisure. When work is neglected, freedom disappears under pressure and anxiety. Discipline creates choice by preventing last-minute crises, allowing students to enjoy life without constant stress.

Question 5: How do study habits function as training for life beyond academics?
Answer: The chapter stresses that the habits formed through consistent study—time management, focus, responsibility, reflection, and perseverance—extend far beyond school. These habits shape how individuals approach work, relationships, and long-term goals. Education, therefore, is not only about acquiring knowledge, but about developing character. Keeping up with studies becomes preparation for managing life itself with confidence and clarity.

The End